Agricultural S[cience]

A Junior Secondary Course for the Caribbean

450

Ian Elliott

Orville Wolsey

Advisors: **Clive Hammans** (Barbados)
Emmanuel Roberts (Grenada)
Vincent Morain (Grenada)

LONGMAN

Publishing for the Caribbean

Pearson Education Limited
Edinburgh Gate
Harlow
Essex
CM20 2JE, England
and Associated Companies throughout the world.

First published 1995
Sixth impression 2001

Set in Palatino 11/13pt

Printed in Malaysia, PA

ISBN 0 582 21090 9

About the authors:

Ian Elliott is an experienced teacher of agriculture and author of many successful textbooks on agriculture for schools and teachers. He has worked with the UK Overseas Development Administration as a Senior Inspector and Curriculum Developer for Agriculture in Schools, and has served as a Consultant on Agricultural Education for UNESCO and the EEC.

Orville Wolsey has over twenty years of teaching experience at junior secondary level. He has contributed to syllabus development in agricultural science in Trinidad and Tobago, and is also a practising farmer. He is currently developing the agriculture programme at Corinth Teacher Training College, Southern Trinidad.

ACKNOWLEDGEMENTS

The Publishers are grateful to the following for permission to reproduce photographs in the text:

A-Z Botanical Collection for figs. 1.13a (Matt Johnson), 1.13b, 4.10, 4.12 and 5.6 (Mrs. W. Monks); Heather Angel for fig. 4.8; Biophoto Associates for fig. 4.9; J. Allan Cash Photo Library for figs. 15.6a, 15.6b, 16.7 and 16.8; Caribbean Agricultural Research and Development Institute, Barbados for fig. 1.10, 1.14, 7.1, 12.4b and 13.7; Bruce Coleman Ltd. for figs. 6.3 (Dr. Frieder Sauer) and 13.1d (Rod Williams); Crown Copyright reproduced with permission of the Controller of HMSO for figs. 13.18a and 17.16; Department of Information, Uganda for fig. 3.6; Ian Elliott for figs. 16.13a, 16.13b and 16.13c; M.M. Gittens, St. Lucia for figs. 1.8, 4.11, 8.8, 9.2, 12.4c and 17.19; Griffin and George for fig. 17.17; Robert Harding Picture Library for figs. 1.3(Wally Herbert) and 17.1(Richard Elliott); The Hutchison Library for figs. 1.2 and 1.4 (Anna Tully); Frank Lane Picture Agency for figs. 13.1a, 13.1b, 13.1c, 15.11a, 15.11b, 15.11c and 15.11d; Massey Ferguson for fig. 10.10; National Meteorological Library for fig. 17.9; Negretti Automation Ltd. for fig. 17.12; Novelty Trading Co. Ltd., Jamaica for fig. 12.4a; Planet Earth Pictures/John Lythgoe for fig. 4.7; Roslin Institute, Midlothian for figs. 13.3b and 13.8; David Simson for figs. 1.11 and 1.12; M.G.V. Thompson, Bradfield College for fig. 13.3a, Trinidad and Tobago High Commission, London for fig. 2.6; Professor Fergus Wilson for fig. 16.18 and Zambia Information Services for fig. 13.9.

The cover photographs have been kindly supplied by Stephen Benson (top left, bottom left and centre), Holt Studios International (top right) and Trinidad & Tobago High Commission (bottom right).

Figs. 17.2 and 17.4a and cover (centre right) are © Longman Group

We are grateful to the following copyright holders for permission to reproduce poems:

The author's agent for 'The song of the Battery Hen' by Edwin Brock; Mrs D. Carberry for 'Nature' by the late H. D. Carberry; Mrs E. Collymore for 'The Bee' and 'The Pig' by the late Frank Collymore; the author's agent for 'June Bug' by Edward Lucie Smith. We have been unable to trace the copyright holder of the poem 'The Eleventh Commandment' by W. Loudermilk (translator Salazar) and would appreciate any information that would enable us to do so.

The author, Orville Wolsey, would like to thank his severest critics and greatest support – his family. He would also like to thank colleagues for their encouragement before and during the project, and all those who contributed information or advice during the research phase.

CONTENTS

USING THIS BOOK

This book aims to tell you *what to do* in agriculture and *why*.

Watching how plants and animals grow is fascinating, especially when they are your own. The book shows you how to run small farm projects and look after your crops and livestock. There are step by step instructions for all the things you need to do.

Then the book explains for each topic *the reasons why* tasks are done in a certain way. Farmers need a good reason for what they do. They don't like wasting their time and effort, and neither do you!

Whenever you do anything in agriculture, start by thinking like a farmer. Ask:

'What is the *best way* of doing this?'
'*Why* should we do it like this?'

When you know the answers to those two questions, you will have learned something useful.

Learning by doing

'Learning by doing' is a good way to learn, because you gain skill and knowledge at the same time. This is how all practical subjects are learned, e.g. home economics is learned in the kitchen, technical subjects in the workshop, and agriculture in the garden or on the farm.

A motto that sometimes appears on classroom walls says:

> What I hear, I forget
> What I see, I remember
> What I do, I understand.

This is why in your agriculture course, and in this book, each new topic leads into activities, observations and experiments. These are followed up by the reasons and explanations.

Science and agriculture

Agriculture is an 'applied science'. It is a science that is useful in daily life and work. We need the science to help us with the practice.

Knowing about the biology of plants and animals helps us to understand their needs and look after them properly. Knowing about the physics and chemistry of soil helps us to make the garden a better place for crops to grow.

When you learn agriculture, you learn a lot of useful science at the same time. Your skills and your understanding grow together. This is the kind of knowledge that will be of real practical use to you long after you have left school.

STUDY SKILLS

Here are some hints to help you in your work at school. Everyone can learn better and quicker when they become good at:

1 Observing
2 Outlining
3 Questioning
4 Measuring and Recording
5 Reporting (oral or written)
6 Reading

These are the 'study skills' or ways of helping yourself to learn.

1 Observing

This means using all your senses, (sight, hearing, sense of smell, sense of touch). Keep your eyes and ears open, and miss nothing. Farmers, of all people, have to be observant. They often spend time just standing still and looking at their crops and animals to see if they are normal and healthy, or if anything needs to be done about them. They observe carefully before deciding what to do.

2 Outlining

'Outlines' are ways of showing the essential points of a subject. For example, in a text book you often see certain things that stand out, like:

Headings, **Bold print**,
CAPITALIZATION, <u>Underlining</u>, *Italics*,
First sentences in paragraphs.

Look at chapter 1, and find examples of these. They are meant to draw attention to what is important. They help you to check through a chapter quickly to see what it is about. Once you can see the outline, you can grasp the details more easily.

3 Questioning

If you read or hear something you cannot understand, ask your teacher about it. Teachers like answering questions! It helps them to help you, and when you ask your question, you are likely to remember the answer.

4 Measuring and Recording

Measuring and recording are essential in science. You often need to record quantities, times, rates of growth and weights of materials and supplies. Good record-keeping is an essential study skill, and it helps you to remember.

Aim to set out tables, graphs, labelled diagrams and written notes neatly and clearly. Practise your skill each day as you do your work.

5 Reporting

This means describing a topic you have investigated or an experiment you have carried out so that others can understand it. Here are some suggestions on making a report:

- Give the facts briefly and clearly.
- Define any new terms.
- Let each sentence describe a single idea.
- Let each paragraph be a series of connected ideas.
- Go through the topic step by step. Complete one step before moving on to another.

- At the end, summarize what you have said.

6 Reading

You can find out practically anything from a library if you know how to use the books. Here are some tips:

Finding the information you want

Usually you do not need to read a whole text book to find the facts you need. You just need a few pages at a time.

Look in the contents at the front, and the index at the back to find the topic you want. Read what you need to read. It might take you only ten minutes. Then put the book away. You could check what several books have to say about your topic, and look at their pictures and diagrams, in quite a short time.

Understanding

When you understand something, you usually don't forget it. So always read to understand, rather than to memorize.

Never attempt to memorize things you do *not* understand. If something in a book is hard to follow, ask your teacher to explain it.

Revision

When you hear the same thing repeated several times, it stays in your mind. Think of television advertisements. They make you remember things you do not really want to know, just by repeating the message every night for a week.

Revision is like this. If you forget something, read it again. Keep having another look, and another, until you know the subject well. It takes time, and you have to keep at it, but this is the easiest and most natural way to learn.

You can revise frequently without worrying too much. Just keep 'repeating the message' and you may be surprised how well you remember.

Study well!

AGRICULTURE IN THE CARIBBEAN

When you have read this chapter you will be able to:

- *explain what is meant by agriculture.*
- *describe how agriculture developed from hunting and gathering to settled farming.*
- *describe the different types of agriculture being practised in your own area.*
- *list the different agricultural activities in the Caribbean countries.*

Fig. 1.1 *Parts of the agriculture industry.*

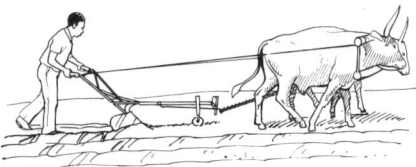

a) Cultivating the soil

b) Weeding a crop

Our word 'agriculture' is actually made up of *two* words taken from the language of Latin. The first part of the word, **'agri-'**, comes from the Latin **'agrarius'**, meaning 'of the land'; and the second part, **'-culture'**, comes from the Latin **'cultura',** meaning 'to till or cultivate'.

Today, agriculture is a major industry with many branches.

Exercise

Look at the pictures and try to describe in your own words what you mean by agriculture. For example, can you say:

- *what farmers do with the soil?*
- *what they do with plants?*
- *what they do with animals?*
- *how they sell their products?*
- *what happens to these products when they leave the farm?*

Each of these activities is a part of the agriculture industry. It includes everything from food production to

c) Harvesting

d) Animal husbandry

e) Marketing

f) Processing food products

Fig. 1.2 *Hunter gatherers.*

the processing and sale of farm produce. We must not forget that the industry has a business side to it. Farmers grow what people want to buy, and they hope to make a living from the sale of their produce. The industry is therefore organized and controlled by the demands of the market.

The history of agriculture

Hunter gatherers

Early man did not practise agriculture as we know it. The first people are believed to have lived by **hunting** animals and **gathering** wild fruits, roots and berries. This is why they have been given the name of **hunter gatherers**.

Nomadic herdsmen

Much later, people began to tame animals. Some followed grazing herds which moved from place to place to find food. People who follow wandering herds are called **nomads**, and even today there are nomadic herdsmen, such as the Lapps of northern Europe, who follow their reindeer herds. Some drawings made on rocks still exist showing Amerindian animals which were herded by the people of that time.

It takes time and patience to tame animals, so that they are not afraid to stay near people, but eventually these herdsmen learned to **domesticate** animals. They chose the quiet, less nervous animals, and bred from them, and began to keep herds near to home.

The beginnings of farming

People began a primitive kind of farming when they started deliberately to plant the seeds of plants that they knew were useful for food. They began to grow **crops** and look after them. Can you imagine what it must have been like to do this for the first time? What problems might these early farmers have met?

Often the people practised **shifting cultivation**. They cleared an area of land by burning down trees and bush, and farmed there until the soil was no longer fertile. Then, after three to five years, they would move to

Fig. 1.3 Nomadic Lapp herdsman with reindeer herds.

Fig. 1.4 Forest being cleared for cultivation.

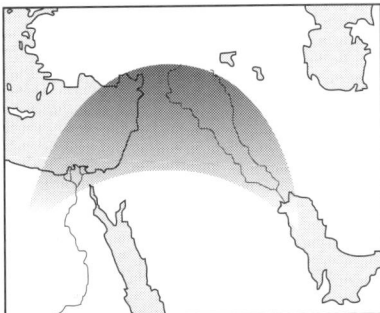

Fig. 1.5 The 'fertile crescent'.

another piece of land. This form of agriculture tends to destroy natural forest and reduce the fertility of the soil, and today we would think it wasteful and inefficient.

There are still places in the world where ancient rainforests are burned to make way for farming. Do you think this is a good idea? Can you give some reasons for your answer?

Settled agriculture

Eventually, a more settled form of agriculture became the normal pattern, and people stayed in one place, farming the land and raising their animals. Because they stayed in one place, they could build, and accumulate possessions, and it was this settled way of life that made the beginning of civilization possible. People began to have time to think about things other than where their next meal was coming from.

Historians believe that agriculture developed in the Near East about 4000 B.C., in the area known as the **fertile crescent**. However, recent research has shown that cultivation of crops and domestication of animals took place even earlier, (between 7000 and 9000 B.C.), in several widely separated places, and that plants were probably being cultivated in the New World.

Agriculture in the Caribbean

There is evidence that the early inhabitants of the Caribbean region practised agriculture. These were the **Amerindians**, (Arawaks and Caribs). They grew crops such as cassava (which they called manioc), tobacco (from which Tobago derived its name), maize, sweet potato, chilli peppers, and cotton (from which they made cloth and hammocks); as well as a variety of tropical fruits such as guava, custard apple, paw paw and pineapple.

The Amerindians developed a simple technology for planting, using the 'digging stick', and making mounds of earth to protect roots of plants in the dry season. They practised composting, and even made simple machines for processing their foods, such as graters and juice squeezers, and they made large flat ovens in which they baked cassava cakes. Besides cassava cakes, they cooked casareep and pepperpot, and brewed beer from maize.

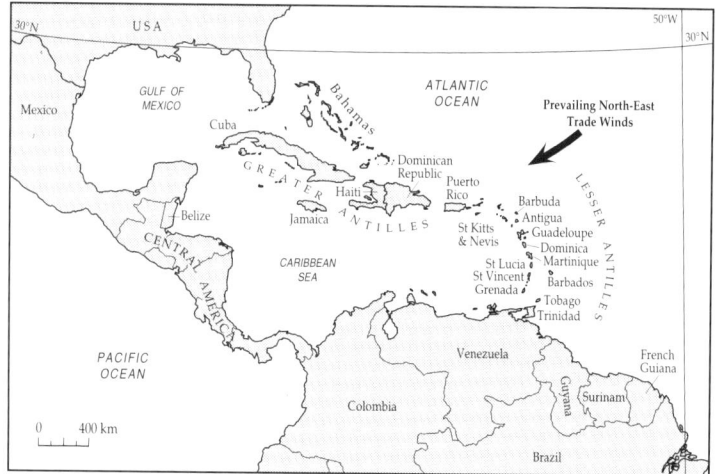

Fig. 1.6 The Caribbean Islands in relation to Mexico, Central and South America.

Fig. 1.7 Amerindian people fishing.

The Amerindians also practised fishing. They constructed dugout canoes and pirogues. Amerindian women reared poultry, so they had eggs and meat from their birds.

You will probably know from your history lessons that the Amerindians were displaced by European invaders, who established great plantations for growing sugar cane and tobacco. The plantations were run using slaves brought over in large numbers mostly from African countries. This system of forced labour came to an end eventually when the slaves were emancipated. Many of the old plantations were abandoned and gave way to smallholdings. Here some of the freed slaves established themselves, growing food both for their own families and for selling in the local markets.

You now live in the very same area where all this happened. Imagine what life would have been like in the Caribbean countries before the coming of Columbus. Would you like to go back to the old way of life where people subsisted by practising simple agriculture and fishing? Even today, the way of life of many people follows this same pattern.

Crops

You are fortunate to live in a part of the world where a wide variety of tropical and sub-tropical plants can be grown. Crop plants from more temperate lands have also been introduced successfully. Different crops are grown in different parts of the region.

Can you say which crops are grown in *your* own country, and why they have been chosen? You will find a list to help you on page 51. Copy and fill in this table:

Crops grown in your country	Reasons for choosing these crops
1	
2	
3	
4	
5	

Fig. 1.8 A fish market.

Fig. 1.9 A farm visit.

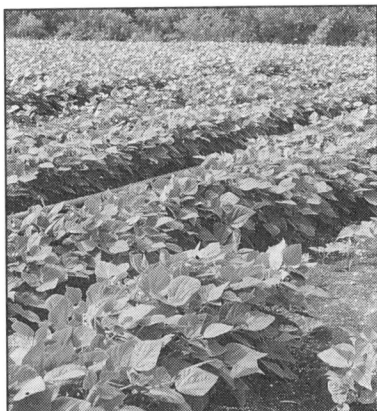

Fig. 1.10 Crop production.

Farm animals

There are also a lot of different kinds, of farm animals kept throughout the Caribbean. Here are some of them:

cattle	sheep	pigs	rabbits
buffalo	goats	chickens	guinea fowl

Can you say which ones are kept in your own country? Copy and fill in this table:

Animals kept in your country	Reasons for keeping these animals
1	
2	
3	

4	
5	

Fig. 1.11 Canning grapefruit.

Fig. 1.12 An ornamental garden.

Fig. 1.13 Products of tropical agriculture.

Fishing

Fishing (Fig. 1.8) is still an important part of the way of life in many areas. Can you list some kinds of fish and other seafoods taken from the sea near to where you live?

PRACTICAL WORK
A farm visit

1 Ask your teacher to take you on a visit to the school farm or to a nearby farm.

2 Before you go, prepare some questions you would like to ask. When you get there, try to find out as much as you can. Take a pencil and notebook with you and note down what you see.

3 Back in the classroom again, write a report on your visit.

4 Walk around your neighbourhood and notice the many activities related to agriculture taking place there.

5 Look in your neighbourhood for:

Home gardens kept for growing food for the family, and for making pleasant surroundings for the home.

Subsistence farming where a person has a job outside of farming, but practises small scale mixed farming after work.

Small farming carried out by full-time farmers who have between one and two hectares of land. Most farmers in island communities are in this category. They produce food

Fig. 1.14 Agricultural research.

Fig. 1.15 Agricultural activities in Trinidad, Tobago, St Lucia and Guyana (see also page 8).

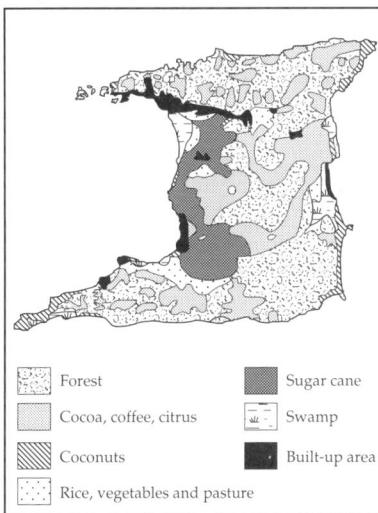

Forest	Sugar cane
Cocoa, coffee, citrus	Swamp
Coconuts	Built-up area
Rice, vegetables and pasture	

Trinidad

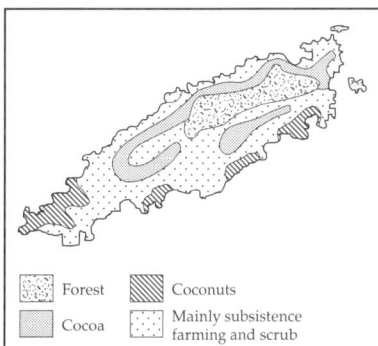

Forest	Coconuts
Cocoa	Mainly subsistence farming and scrub

Tobago

crops, vegetables and animal protein for the local market.

Estates large farms owned by individuals, families or companies. They usually produce plantation crops for export.

6 Draw a map of your neighbourhood and record these activities on your map. Name the crops grown by those who practise agriculture in each of these different ways.

Agricultural activities

We are surrounded by agriculture, but as we have already seen, the agriculture industry includes more than just farming itself. Some activities take place on the farm while others take place off the farm. Figs. 1.10-1.14 show some activities associated with agriculture.

Crop production.

Many different types of crops are grown in fields and gardens. Can you name those shown in Fig. 1.10?

Animal production.

Farmers raise young animals to replace the ones that are sold. They must know how to look after these young animals.

Processing of agricultural products.

Many of the foods we eat have been **processed** in some way. They may have been canned or dried, baked or frozen. Can you think of some foods which have been treated in these ways?

Marketing of agricultural products.

Farm produce is sold in the market. **Marketing** is therefore part of the agriculture industry.

Ornamental horticulture.

We like our surroundings to be pleasant. Beautiful flowers and foliage plants are grown in parks and gardens.

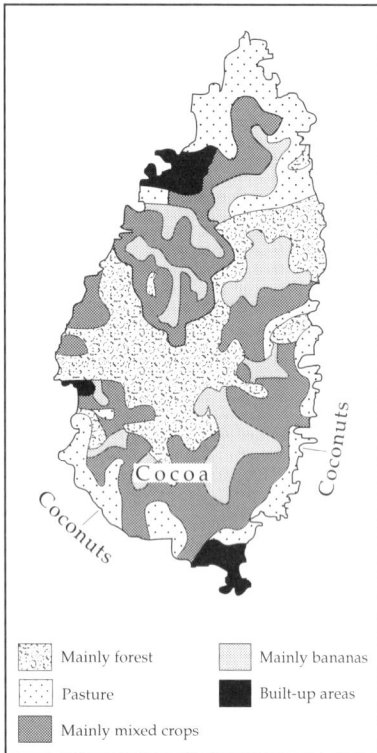

St. Lucia

Guyana

Tropical exotics and niche market crops.

Some fruits and flowers grow in tropical areas only. They can be sold for a high price. Examples are hot peppers, carambola, aloe clera, orchids, heliconias, shadon benny, etc.

Agricultural research.

Agriculture is constantly being improved through **research**. On special farms called **research stations** experiments and trials are carried out to develop new techniques, and to develop more productive strains of plants and animals.

Exercise

1 For each of the agricultural activities shown in the pictures (Figs. 1.10–1.14), write a short description of what is happening. For example, can you say:

- *what kind of tools or machinery are required for this activity or producing these products?*
- *how many people might be needed to do the work?*
- *what special skills they might need?*

2 The maps in Fig. 1.15 (pages 7 and 8) show how land is used for various agricultural activities in several countries. Draw a large outline map of your own country. You can use your atlas to help you. Show on your map the areas where different agricultural activities take place.

Questions

1 What is agriculture?

2 List the different types of agricultural activities which take place in your school or home district.

3 Write a note to a friend entitled 'Agriculture then and now' (comparing agriculture a long time ago with agriculture today). You could draw a comic strip or write a short play if you prefer.

4 Explain the following terms:

(a) subsistence farming

(b) home gardening

(c) small farming

(d) estates

2

AGRICULTURE – THE BASIC INDUSTRY

When you have read this chapter you will be able to:

- *explain how agriculture can provide the people of the country with:*

    - *food*

    - *clothing*

    - *employment*

    - *money*

    - *trade.*

- *say why it is worth learning about agriculture in school.*

- *describe an agro-industry close to your school.*

- *list the products of this industry and say where they are sold.*

Agriculture is the main industry in very many countries. It is the biggest industry in the Caribbean, as you already know. In studying agriculture, we must first ask:

– Why do people need agriculture?

– How can agriculture help us?

In this chapter you will find some answers to these questions.

Things provided by agriculture

Food

Agriculture provides food. People have to eat – and food comes from farming. Whether they grow their own food, or buy it in the market, someone somewhere must produce it – a lot of it – to provide for the whole population.

People also need a good *variety* of different kinds of food. This gives them a **balanced diet**, to keep them alive and healthy.

It is obvious that without agriculture there would be nothing much to eat. There would be no domestic animals or cultivated crops, no vegetables and no fruits. Very soon there would be no people! Agriculture is sometimes called 'the basic industry'. It supports all our lives, and the better our agriculture is, the better fed and the healthier we are.

Exercise

Make a list of some of the crops cultivated around your home or school which are grown for food.

Employment

Agriculture gives employment. Many people living in the Caribbean work directly in agriculture, growing crops and raising animals. Others work to supply farmers with what they need – seeds, fertilizers, tools and machines. Others handle, store and transport farm produce, and factory workers process agricultural products. In fact there are very many jobs in and around agriculture that do not involve actual farming. Researchers, teachers and trainers are all working to help agriculture succeed, even though they are not farmers themselves.

Exercise

Make a list of as many jobs as you can think of that have something to do with agriculture.

Money

Money can be made from agriculture. Farmers start with the soil, seeds, fertilizers, animals, and tools. They grow their crops and rear their animals. They then sell the products from these activities in the market.

Fig. 2.1 Good food makes you strong and healthy.

Fig. 2.2 Growing money.

Fig. 2.3 Farming is a way of making money.

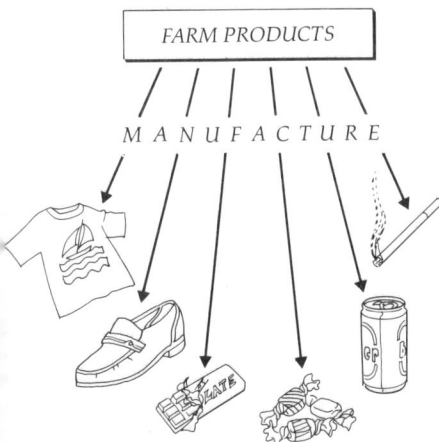

Fig. 2.4 Manufactured goods made from farm products.

Notice what is happening. Starting with things of *little value*, farmers produce things that have a much *greater value*. Farmers 'create wealth'. They grow money!

Trade

Agriculture makes trade possible. Because they have something of value to sell, farmers can trade in the market. They can use the income from their own produce to buy other things they might need. That is the purpose of markets, as you will know if you have ever been to a market and seen the buying and selling that goes on.

It is important to grasp that what happens in local markets is really just the same as what happens in trade between the countries of the world. Different countries trade with each other just as people do in the market. A country may *sell* some things that it produces, and then *buy* things that it *cannot produce* from other countries.

Climate has a lot to do with it. For example, in hot countries crops like coffee and cotton can be grown which cannot be grown in more temperate zones. Other countries may produce cars and tractors, farm tools and machinery – things that the agricultural countries do not have. So different countries can make **trade agreements**, and sell to each other.

Raw materials

Agriculture provides raw materials for industry. Food is not the only product of farming. Agriculture produces things that are then manufactured in various ways. We say that the **raw materials** come from agriculture to produce **manufactured goods** such as clothes and leather items, chocolate and sweets, beer and cigarettes.

Exercise

1 *Make a list of manufactured goods made from farm products. On your list, write the names of the raw materials supplied from farming to make each of these goods.*

2 *Can you list the main exports of your country? Are they mainly agricultural raw materials, or mainly manufactured goods, or mainly fuel? Use your atlas to help you.*

3 *Can you list the main imports of your country?*

Because agriculture is the basic industry, concerned with the production of food and raw materials, a farmer is an important member of any community. He is a **primary producer**. If it were not for people like him, there would not only be no food, but there might be no wealth and no trade either! Every government tries to encourage agricultural production, because so much of the prosperity of a country can depend on it.

Fig. 2.5 Something to think about!

'Why should we learn about agriculture?'

There are very good reasons for learning about agriculture in school. Here are some of them:

- Food prices keep rising. If you know how to produce even *some* of your own food, this will help to improve your nutrition and save you money.
- If you can produce some *extra* food to sell, you may decide to supplement your family income.
- Growing vegetables in your own garden is a useful leisure activity for the whole family.

'Suppose I get a job outside of agriculture, what then?'

Even if you do not want a job in agriculture when you   leave school, it will still help you to know about it. Everybody should know what is going on in their own country. Think about it:

- Whatever you do later on in your life, you will be able to understand what the farmers are doing and how an important industry in your country operates.
- Whatever job you do, you are likely to meet people who work in agriculture.
- Even traders, bankers, and business people deal with people who work in agriculture in some way (farmers, agricultural merchants, shopkeepers selling farm produce etc.).

Agriculture is one of the most useful subjects you could learn.

Fig. 2.6 Agro-industry.

PRACTICAL WORK
Agro-industries close to home

If you are near a sugar mill, or a fruit growing plantation, or an abattoir, you might be able to go and see it. Most people are pleased when students are interested in what they are doing, and are happy to show them around.

Just as you did when you visited a farm, prepare some questions in your notebook to ask when you get there. Here are some questions to give you the idea. Copy these questions out, and add some of your own, so that you do not forget what to ask. You might think of other questions when you arrive:

- *What farm products does this agro-industry process?*

- *How are they taken from the farm to the place where they are processed?*

- *How is the farm produce processed when it arrives? (Leave a lot of space for this answer.)*

- *What are the processed products of the industry?*

- *Where are they sold?*

- *Who buys them?*

- *How are these processed products transported to market?*

- *How many people are employed in the industry?*

- *Do you consider this industry important to the community and the country? Can you explain how or why?*

- *Are there jobs for school leavers in the industry?*

- *Do you provide training for young people?*

- *What kind of training do you provide?*

As always on a school visit, keep your eyes and ears open at all times! Remember to behave well, be polite and say 'thank you' at the end of your visit. Then you might get asked back again.

Questions

1 List the basic needs of man that are supplied by agriculture.

2 Make a list of some jobs related to agriculture which do not involve primary production.

3 Explain the statement, 'Agriculture is a source of income for the farmer and a source of revenue for a country'.

4 Match the manufactured products to the new materials (farm products) from which they were made.

Raw materials

sugar cane
milk
beef
eggs
chicken

Manufactured products

granulated sugar
condensed milk
corned beef
beef salami sausages
frosted cake sponge
ice cream
chicken vienna sausage
candy
rice cakes
UHT milk
milk chocolate
hamburger
shirt
eggnog

5 Write a letter to a friend to convince him or her why you think he or she should study agriculture.

3

PLANTS AND AGRICULTURE

When you have read this chapter you will be able to:

- *explain what plants need to make them grow.*
- *explain why people and animals are dependent on plants.*
- *describe how plants provide us with:*
    - *food*
    - *clothing*
    - *useful materials.*
- *explain the importance of looking after plants and their environment.*

Fig. 3.1 What is this plant doing?

If you look at a plant, you cannot see it doing anything. It keeps quite still, unless the wind blows it, and it seems to make no response to its surroundings at all!

We know that plants are alive, because they slowly grow bigger and change. After a long time they produce flowers and fruits. They must be feeding in some way to make them grow like this.

We also know that plants do need certain things to make them grow – things like water, light, and warmth. But at first sight you cannot tell *why* they need these things, or how they use them. It is as though plants were aliens from another planet, whose mysterious ways of living are very unlike our own.

If we want to help our crop plants to grow well, we really must learn, if we can, how they manage to live in their strange, vegetative way. If we *can* understand their way of living, we may be able to provide them with the things they need.

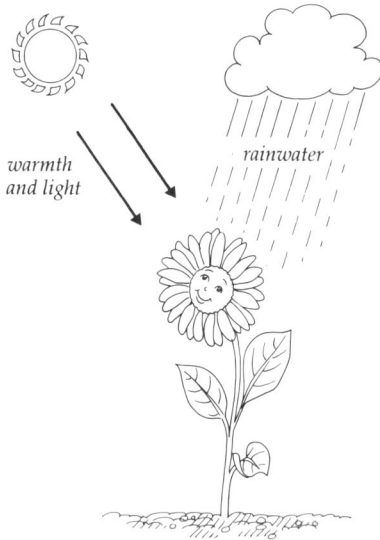

Fig. 3.2 A plant needs water, light, warmth and air.

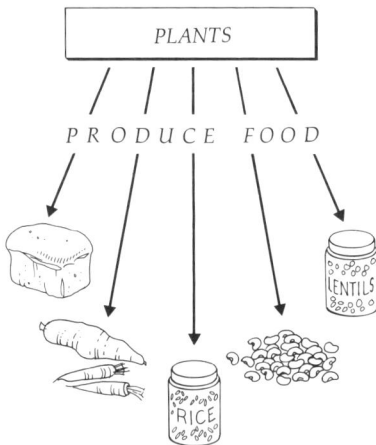

Fig. 3.3 Foods from plants.

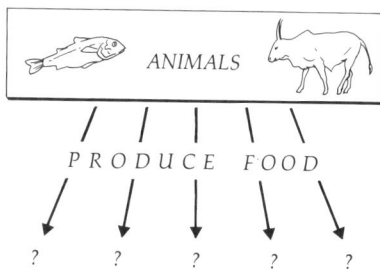

Fig. 3.4 Foods from animals. What do these animals eat?

In later chapters you will be learning about the **life processes** of plants.

Plants and people

It is an interesting thought that people need plants more than plants need people! Plants could live perfectly well without us, but we certainly could not live without them. They supply us with so many of the things we need every day – our food, our clothing, all sorts of useful materials, and even the oxygen in the air we breathe.

Uses of plants

1 Food from plants

Exercise

1 *Make a list of ten foods you eat regularly that come from plants. This should not be difficult. Here are three to start your list . . . lentils, beans, rice.*

2 *For each food you have named, can you say what part of the plant you are eating? Is it the stem, or the root, or the seed?*

3 *Now see if you can make a list of five foods you like to eat that come from animals.*

4 *For each of these foods, can you say what kind of plants the animal usually feeds on?*

Animals may feed on other animals. Often a **food chain** exists, when animals feed upon each other in sequence; but if you trace back the food chain to the beginning, you will find that it starts with a green plant of some sort. Here are examples of three food chains:

grass $\longrightarrow$ goat $\longrightarrow$ man
cabbage $\longrightarrow$ caterpillar $\longrightarrow$ bird
algae $\longrightarrow$ water flea $\longrightarrow$ fish

Note that grass, cabbage and algae are all green plants.

All life depends ultimately on the green plants. There could be no life on earth as we know it without them. The reason is that green plants *make* the nutrients that are present in the things we eat – foods like **carbohydrates**, **proteins**, **fats** and **vitamins**. You may

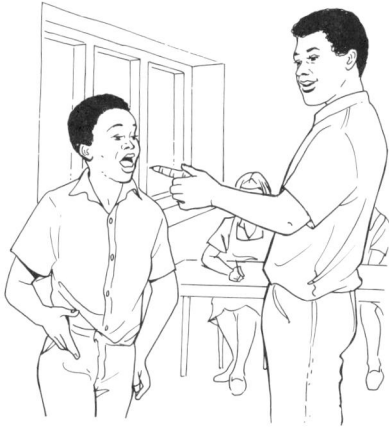

Fig. 3.5 Whose food is it?

have learned about these in Biology, Integrated Science or Home Economics lessons. Animals and people *must* have these foods to survive, yet they cannot make them for themselves. They have to get them from plants.

We should realize also that plants make the food for *themselves*, and not for us! They store food in their bodies, so that they can use it later. For example, a wheat plant stores food in its seeds, so that when the seeds germinate they will have enough food to give them the energy needed to start growth. A person may then come along and take the seeds to make bread, but this was not the plant's idea. It did not store the food for *him* or *her*!

Or, to take another example, a carrot plant stores food in its root, so that next year it can grow an **inflorescence** (flower), and produce seeds. But we dig up carrots and eat them before they can complete their life cycle. Again, the plant makes food for *itself*, but we take the food and eat it.

Farmers deliberately grow plants that are good at storing food in their roots or seeds. These are our crop plants.

2 Clothes from plants

The cotton plant supplies fibre that can be made into cloth. Cotton cloth is the ideal fabric to wear in hot climates, because it protects us from the sun without making us feel too hot, and absorbs the moisture from our bodies.

Fig. 3.6 Cotton fibres can be spun into yarn.

You will probably know that cotton comes from the lint that surrounds the seeds of the cotton plant. The plant produces the lint to help it spread its seeds about. If the plant were left alone, the wind would eventually blow away the soft, light lint and carry the seeds with it. The plant relies on the wind to disperse its seeds in this way over a wide area.

People have learned to separate the fibre from the seed, spin it into a yarn, and weave the yarn into cloth. Again, we are taking what the plant does for its *own* purposes and turning it to our advantage. The plant does not want to provide us with nice clothes! It is just trying to disperse its seeds.

Artificial fabrics can now be produced from **synthetic**, (man-made) fibres such as rayon, nylon and acrylic fibre. These make very good clothes, but many people still prefer to wear cotton because it feels more comfortable. Cotton fibres are sometimes mixed with synthetic ones to make clothes that are both comfortable and hard wearing.

3 Useful materials from plants

There are so many useful materials from plants that a whole book could be written about the plant products we use every day.

Exercise

Make a list of as many plant products as you can, and describe the use or uses to which they are put. Here are a few suggestions to start your list:

Plant product	Use
Wood	buildings
	furniture
Hemp fibre	rope and string
Oil	margarine
Sugar cane	sugar

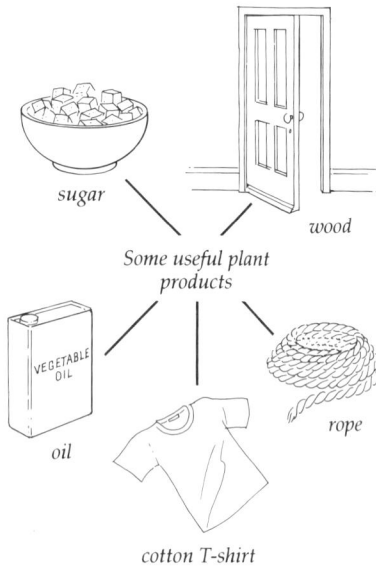

Fig. 3.7 Some plant products.

Plants and the environment

If it were not for plants, our environment would soon be unable to support our lives.

These are some of the ways in which plants make our environment better for us:

1 Plants recycle waste materials.

When plants or animals die, they rot in the soil, and are used as food by living plants.

2 They help to control soil erosion.

Plants protect the soil from falling rain with their leaves, and they hold the soil together with their roots, preventing it from being washed away.

3 They recycle gases in the air.

The gas **oxygen** from the air is being used up all the time in the respiration of people and animals (and also of plants), and in the burning of fuels like wood, coal and petrol. The gas **carbon dioxide** is produced as a result of these chemical reactions, and it begins to build up in the atmosphere. Some people worry that this could cause **global warming** – a rise in the temperature of the earth's atmosphere.

Green plants, in the presence of sunlight, *use up* **carbon dioxide** in large quantities, and *produce* **oxygen** again. They do this by a process called **photosynthesis**, which will be explained later. We simply could not survive without the oxygen that plants release into the atmosphere every day.

4 Plants collect dust from the air.

Dust sticks on to their moist leaves. This helps to clean the air of dust pollution. Trees are planted near to factories or highways to help keep the air clean. A good example of this is the cement factory at Claxton Bay in Trinidad, where the trees help to reduce air pollution. Can you think of any examples near you?

5 Plants absorb noise.

This is why trees, shrubs and lawns are placed between buildings such as schools. They help to reduce noise levels.

6 Plants act as windbreaks.

Trees and hedges are often planted along highways and around fields and gardens to reduce the force of the wind.

Looking after plants

How would we get on without plants? You can see the answer to that question – we could not survive at all! If we are wise, we will always be careful never to do anything to seriously harm plants, or spoil their chances of survival. If we harm the plants, we can only harm ourselves. If we look after them, they will look after us.

Farmers, of all people, know that it pays to look after plants. They always try to protect the soil and the environment where plants live and grow.

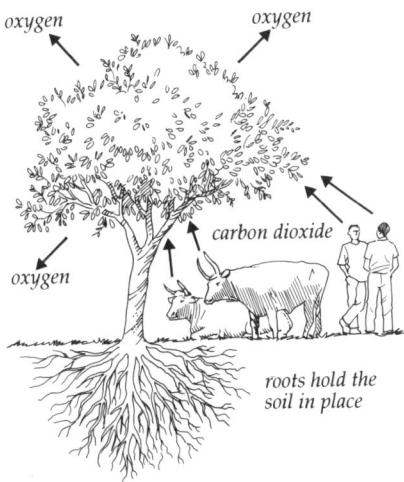

Fig. 3.8 Note how plants protect the soil from erosion and renew the oxygen supply.

Questions

1 (a) Name three things a plant needs from its surroundings.

   (b) How does each of these things help the plant to grow?

2 'If there were no plants there would be no animals and no people'. Explain this statement.

3 Draw a diagram of a food chain and explain it.

4 List ten useful plant products. Explain the uses of each product. You may arrange your answer in two columns, as shown:

   Plant product   **Use of the product**

5 Explain three ways in which plants protect the environment.

4

PARTS OF THE PLANT

When you have read this chapter you will be able to:

- *draw a fully labelled diagram of a flowering plant.*
- *describe the functions of the:*
    - *root*
    - *stem*
    - *leaves.*
- *make collections of pressed plants, and pressed tree leaves.*

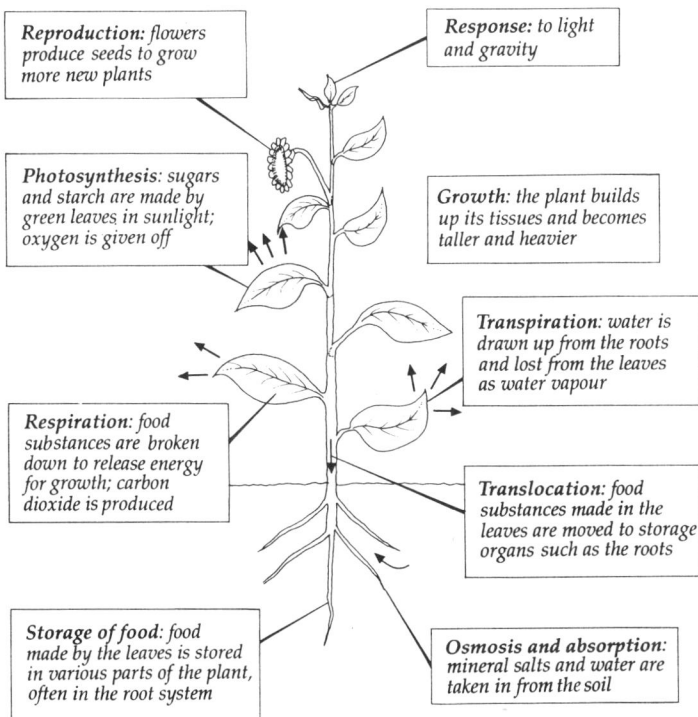

Reproduction: *flowers produce seeds to grow more new plants*

Response: *to light and gravity*

Photosynthesis: *sugars and starch are made by green leaves in sunlight; oxygen is given off*

Growth: *the plant builds up its tissues and becomes taller and heavier*

Transpiration: *water is drawn up from the roots and lost from the leaves as water vapour*

Respiration: *food substances are broken down to release energy for growth; carbon dioxide is produced*

Translocation: *food substances made in the leaves are moved to storage organs such as the roots*

Storage of food: *food made by the leaves is stored in various parts of the plant, often in the root system*

Osmosis and absorption: *mineral salts and water are taken in from the soil*

Fig. 4.1 *Parts of a plant and their functions.*

Part of a plant's body, the **root system**, is below the ground in the soil. The other part, the **shoot system**, is above the ground, in the atmosphere. This means that a plant exists in two very different environments: the soil environment and the atmospheric environment. Not surprisingly, the root and shoot systems are very different. Each is adapted to suit its surroundings, and each has its own special functions.

The root and shoot systems join at ground level. Inside the plant is a network of tiny tubes, or **vessels**, for transporting water and substances dissolved in water. Fig. 4.2 shows the

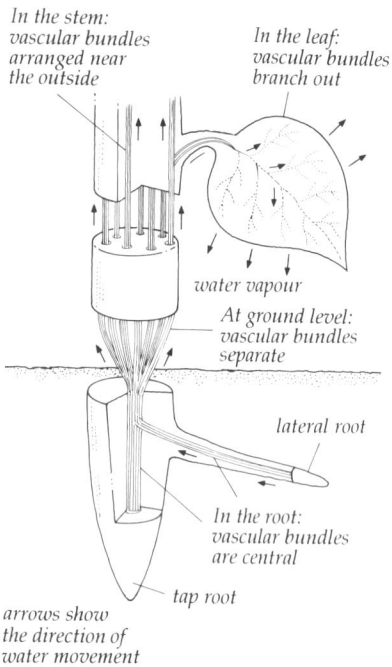

In the stem: vascular bundles arranged near the outside

In the leaf: vascular bundles branch out

water vapour

At ground level: vascular bundles separate

lateral root

In the root: vascular bundles are central

tap root

arrows show the direction of water movement

Fig. 4.2 What a plant is like inside: parts cut away to show vascular bundles.

Fig. 4.3 Looking at plant roots.

layout of this tubular network. It runs right through the root and shoot systems, so that substances in solution can pass from one part of the plant to another.

There are actually two types of tubes. The first, called the **xylem vessels,** carry water and inorganic (mineral) salts in solution, up from the roots through the stem to the leaves. The other type, called the **phloem vessels**, run parallel to the xylem vessels. They carry foods made in the leaves, in solution, to all parts of the plant.

The best way to learn about root and shoot systems is to go out and look at plants around your school. When you are familiar with different types of plants, you can then experiment with some of them to see how they function.

Plant roots

PRACTICAL WORK
Looking at root systems

Dig up a plant gently, and remove the soil carefully, so that you can have a good look at the roots. Try to keep the fine branches of the root system intact.

1 *Go round the school garden looking at different plants. There are narrow-leaved plants, like the grasses, and broad-leaved plants. These include most of the flowers, trees and shrubs.*

Look for some weeds to dig up. You could dig up some garden plants as well, but ask your teacher first. You will need a trowel or a hand fork, and a bucket of water to wash the roots.

2 *Look for two kinds of plant:*

(a) *a broad-leaved weed, e.g. broom weed (Sida acuta), or Crotalaria species (chac chac).*

(b) *a narrow-leaved weed, e.g. savannah or carpet grass (Anoxus compressus) or nutgrass (Cypersus rotundus).*

3 *Dig the plants up gently. The roots are delicate, so lift them very carefully, with the soil still around them.*

4 *Wash the roots in water. Try to avoid damaging the smallest branches of the roots, so that you keep the whole root system intact. Let any excess water dry off.*

Fig. 4.4 *Two types of root system.*

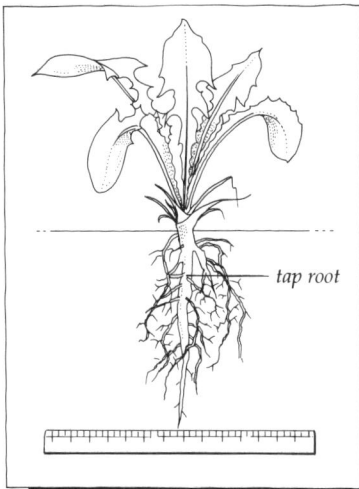

a) A tap root system

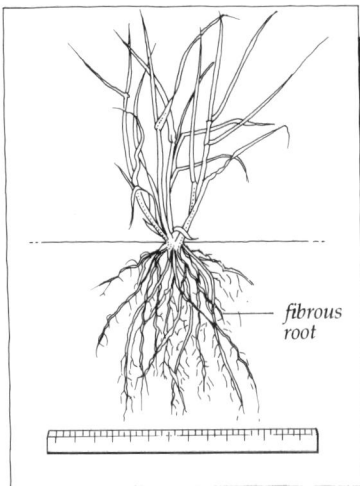

b) A fibrous root system

5   Bring the plants into the classroom. Spread the roots out on separate sheets of paper, so that you can see them. Notice the two types of root systems, and how they branch below the soil. Roots are usually white in colour. They are never green, unlike leaves and stems that contain the green substance called **chlorophyll**. They never have any buds, as stems do.

6   Notice that the plants with broad leaves have a main root, called a **primary root**, growing straight down from the stem. There are **lateral roots**, (side roots), growing out of it, and these branch into smaller and smaller **rootlets.**

7   Plants of the grass family have what are called **fibrous roots**. They have several main roots (**secondary roots**) growing out from the base of the stem. The primary root which grew from the seed dries up after the plant is established.

8   Draw and label both root systems. The pictures will help you.

9   Measure the root systems to find out how deep below the soil surface they were growing, and how far from the centre of the plant they spread out. Write these measurements on your drawings.

10   See how the branching structure of the roots helps them to reach out into a big volume of soil.

11   If you have a lens, or a microscope, magnify the smallest branches of the roots. See how they can get between the soil particles, to reach as much of the soil as possible.

The functions of roots

This means what roots *do*. Roots have four main functions.

1   *Anchorage.* They act as an anchor, holding the plant firmly in place so that it does not fall over, or get blown down by the wind.

2   (a) *Absorption of water.* They take in water from the soil through **root hairs** just behind the root tip. Each root hair is an outgrowth from a single cell, but there are many thousands of

Fig. 4.5

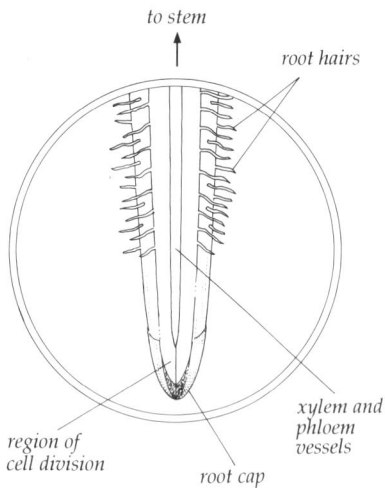

a) A vertical section through a root tip (magnified).

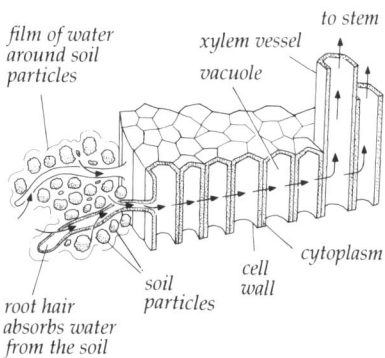

b) A root hair (highly magnified) takes in water and dissolved nutrients; arrows show direction of flow.

them and together they present a very large surface area to the soil. Water is taken in by **osmosis**, a process you may learn about in Integrated Science lessons. Water moves through the semi-permeable membrane of the cell cytoplasm from a weak solution, (the soil solution), to a stronger solution (the cell sap).

After being absorbed, water is passed up through the xylem vessels to the stem and leaves. There it is lost to the air by evaporation. The result is a constant stream of water passing through a plant.

(b)Absorption of nutrients. Plants feed by taking in **inorganic (mineral) salts**. These are substances from the soil that dissolve in water, and are taken in with the water.

3 *Food storage.* Some plants use their roots to store food. These roots are often thick and swollen. Carrot (*Daucus carota*), beet (*Beta vulgaris*), radish (*Raphanus sativus*), and other root vegetables store food in their swollen tap roots.

Other plants such as cassava (*Manihot esculenta*) and sweet potatoes, (*Ipomea batatas*), store food in swollen tubers formed from their roots.

Unusual roots

Plants have become adapted in many strange ways. Here are some unusual structures that have developed from the basic pattern of the root system.

1 *Adventitious roots.* These are roots which grow from parts of the shoot system. For example:

- **aerial root systems** – some plants have specialized roots that do not grow in the soil but are surrounded by air – hence the name 'aerial'. They absorb moisture from the air, e.g. the wild pine (Fig. 4.7).

- **prop roots** of maize (*Zea mays*) that appear from the stem just above ground level and help to keep the stem upright.

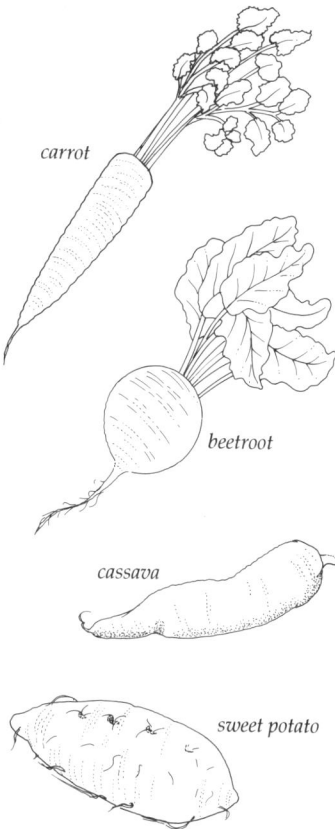

Fig. 4.6 *Roots store food.*

carrot

beetroot

cassava

sweet potato

Fig. 4.7 *Aerial roots of wild pine.*

- **clinging roots** of some orchids (*Bulbophyllum* species) which live on trees; these help the plant to hold on to the tree.
- **water-absorbing roots** are also found on some tree orchids. They can absorb some of the rainwater that runs over them and drips from them.
- **stilt roots** of some mangrove trees (*Rhizophora mucronata*) that help to support the rather weak stems.
- **edge-of-leaf roots** of *Bryophyllum*, ('Wonder of the World'), that appear on the young plants at the edges of the leaves.

2 *Breathing roots.* Some mangrove trees (*Sonneratia alba*) have roots that grow up out of the airless mud to allow oxygen to reach the parts below the surface.

3 *Buttress roots.* The roots of some trees form strong supports around the trunk, e.g. the silk cotton tree (*Ceiba pentandra*). They look rather like the buttresses on some old buildings that help to prop up the walls, and this is where the name comes from (Fig. 4.8).

4 *Legume roots.* Plants of the pea and bean family, called the **legumes**, or (*Leguminosae*), are important in agriculture because they can make use of nitrogen gas from the air (Fig. 4.9). The ability to 'fix' nitrogen from the air is extremely useful, as nitrogen is an important element in plant nutrition. Most other plants cannot use atmospheric nitrogen. Instead, they absorb nitrogen salts from the soil.

Legumes have a micro-organism, *Rhizobium*, living in swellings or **nodules** on their roots. It is actually the *Rhizobium*, not the legume itself, that fixes the nitrogen from the air. The *Rhizobium* uses the nitrogen to build up its own body proteins.

The micro-organisms do not live very long, and when they decompose they release nitrogen salts that the legume plant absorbs through its roots.

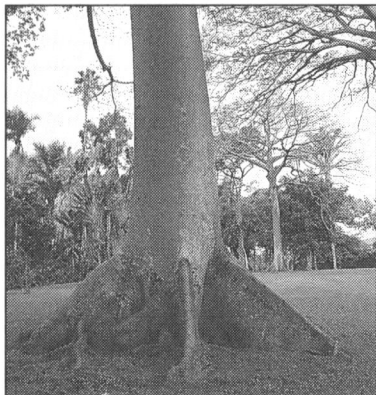

Fig. 4.8 Buttress roots of the silk cotton tree.

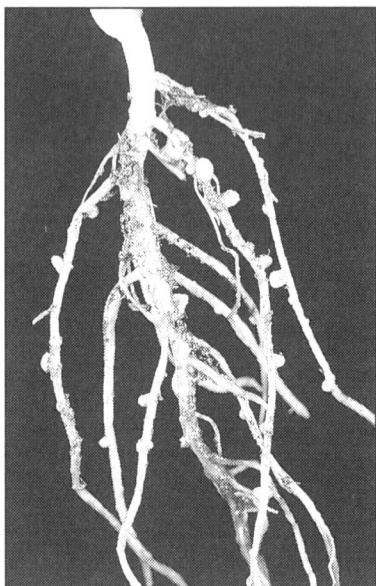

Fig. 4.9 The roots of a legume, showing root nodules.

Fig. 4.10 A mango tree.

Meanwhile, the legumes supply the micro-organisms with carbohydrates. The result of this relationship between the two organisms is that legumes need little or no nitrogenous fertilizer! Eating legumes also gives us protein.

Farmers include legumes in their crop rotation plan, because when the plants eventually die and decay, the nitrogen fixed by them becomes available to the plants that follow them.

Plant shoots

PRACTICAL WORK
Looking at shoot systems

A shoot system is the part of a plant above the ground, made up of the stem, leaves and flowers.

1   *Go out and look at the shoot systems of plants in the area around the school. Take a notebook and pencil to make drawings and notes of what you find.*

2   *Look for examples of the following different kinds of shoot systems:*

* *trees – tall plants, with strong trunks made of wood, and strong branches. These live for many years, e.g. mango (Mangifera indica) (Fig. 4.10).*
* *shrubs – smaller, bushy plants, with several stems made of wood (Fig. 4.11). They live for many years, e.g. croton (Codiaeum variegatum) or hibiscus (Hibiscus rosa-sinensis).*
* *twiners – vines that grow up to reach the light by clinging to other plants and twining round them, e.g.* **runner bean** *(Phaseolus multiflorus) (Fig 4.12).*
* *runners – plants that have stems running over the ground, e.g.* **pumpkin** *(Cucurbita maxima).*
* *climbers – plants with clasping roots that cling on to walls, or other plants, for support, e.g. ivy (Hedera helix).*
* *grasses – plants with narrow straight leaves, growing up from close to the ground.*

Fig. 4.11 Hibiscus.

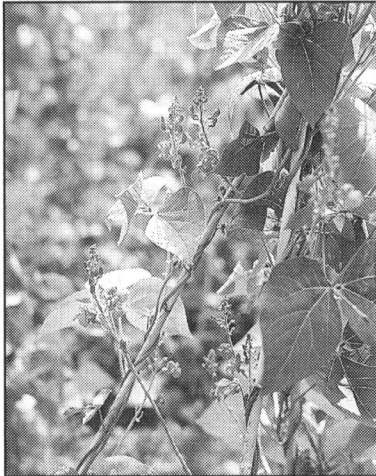

Fig. 4.12 A runner bean twining round a support.

3 *Some stems are* **woody**, *like those of trees and shrubs which live for many years. Other stems are softer, and usually live for just one year. They are called* **herbaceous stems**. *Can you find these different types of stems?*

4 *Notice that shoot systems of the broad-leaved plants, (not the grasses), have:*

- *A stem that grows straight up from the tap root. This results from the growth of the* **terminal bud**.

- *Leaves growing out of the stem along its length. The leaves are spread out to catch as much light as they can.*

- *Buds in the angle, (or* **axil**), *between the leaf and the stem. These are called* **axillary buds**. *They may start to grow, to form a branch.*

- *Branches that have the same structure as the main stem, and bear leaves and buds.*

The buds of most grasses are close to the ground. New stems (fillers) grow from these buds. This is why grasses can be cut or grazed and still recover without being killed.

The shoot systems of most plants grow and branch out from their developing buds. Occasionally, however, new shoots may grow from the root system, e.g. in the ixora plant.

Breadfruit

Have you ever heard of Captain William Bligh?

Go to your school or community library or book club.

See if you can get a copy of *Mutiny on the Bounty* and read the story of how he introduced the breadfruit tree (*Artocarpus altilis*) to the Caribbean.

People did not like it at first. They thought of it as a low status food, but later began to realize how they could cook it and make nice dishes out of it.

Trinidad calypsonian Gorilla in his *Empty Belly* calypso (1938), describes the deep shame brought on by hunger and the necessity of having to eat the breadfruit, considered mainly fit for pigs and '*dem small islands*'.

Fig. 4.13 *Making a collection of pressed plants.*

Many men are working
    for a salary
That can never maintain
    a family
Go to their homes you are
    bound to find
Roast potatoes and saltfish
    cutting shine.
Many other days in the
    time of need
They will crucify a breadfruit
    for a Sunday feed
So you can take a fellow
    like me
Singing this calypso on
    empty belly.

Another calypsonian, Growler, in *The Farmer and the Breadfruit Tree*, describes what the tree has to say about it.

When there was plenty rice
    'bout the place,
They used to watch a breadfruit
    and skin up they face,
They wouldn't agree
    with my intestines
They would cut it up in pieces
    to feed the swine.
But now see them running to the
    market like a racehorse
To try and buy one
    No matter what it cost.'

Collect pieces of shoot systems from different plants and bring them back to the classroom. Make a note of the date and place where you collected them.

Press them between sheets of newspaper, under a stack of heavy books. They will dry out in two to three weeks. It is best to change the newspaper every day to avoid the leaves rotting due to the presence of fungi. When they are dry, you can mount the plant shoots on white paper. Choose the best specimens for your collection.

Write the local name of the plant, (and its generic and specific names if you can), on the page. Also note the type of shoot system it has, and the date and place where it was collected. Your collection of pressed plants will look very good on the classroom wall!

The functions of stems

A plant stem has the following functions:

1 **Connects** the root system to the stem, leaves and flowers of the shoot system.

2 **Supports** all the parts of the shoot system, including leaves, buds, flowers and branches.

3 **Displays** the leaves to maximum exposure of sunlight through its branching habit of growth.

4 **Transports water**, and dissolved nutrients, from the root to the leaves.

5 **Transports food** made in the leaves to other parts of the plant where it is needed.

6 **Storage of food.** Sometimes the centre of the stem is used as a food storage area. Sugar cane (*Sacharrum officinarum*), and cabbage (*Brassica oleracea*) are two examples of such plants.

7 **Perennation.** This means helping the plant to survive from one year to the next in a dormant state. Examples of plants that use modified underground stems as organs of perennation are arrowroot (*Maranta arundinacea*), onion (*Allium sepa*), eddoe (*Colocasia antiquorum*), yam (*Dioscorea alata*), and ginger lily (*Zingeber officinale*). Their underground stems have been modified to form a bulb, a corm, or a rhizome.

Fig. 4.14 *Some plants store food in their stems.*

Plant leaves

PRACTICAL WORK
Looking at leaves

1 *Go out and collect as many different leaves as you can find. Take just one leaf from each plant. Note the*

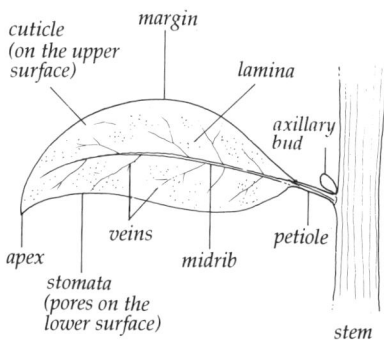

Fig. 4.15 *Parts of a leaf.*

*hibiscus –
example of a
simple leaf*

*pigeon pea –
example of a
compound
trifoliate leaf
(the leaf is in
three parts)*

*tamarind,
sensitive plant–
example of a
compound
parapinnate
leaf (the leaf
has many
parts)*

*cassava –
example of a
simple lobed
leaf*

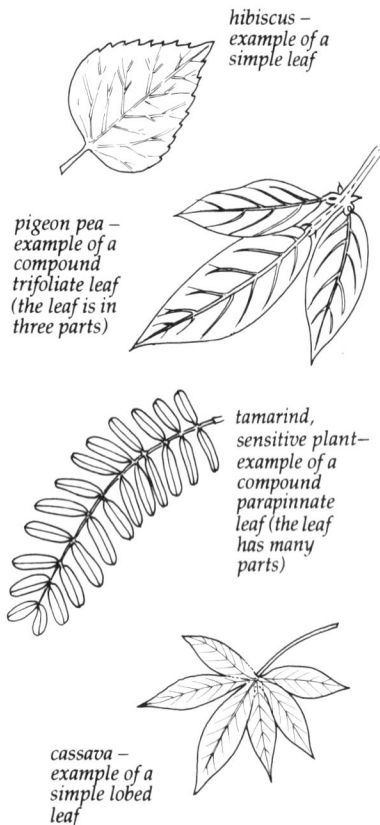

Fig. 4.16 *Some leaf forms.*

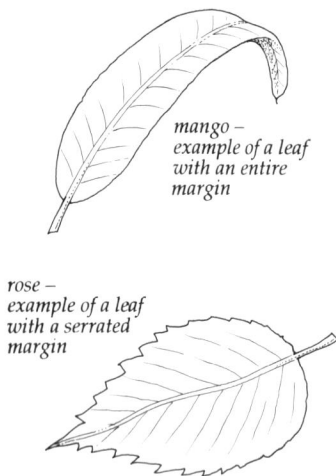

*mango –
example of a leaf
with an entire
margin*

*rose –
example of a leaf
with a serrated
margin*

Fig. 4.17 *Leaf margins.*

*name of the plant if you can. Be careful to take the
whole leaf and not just a part of it. Some leaves have
many parts. Cut off the leaf together with the part of
the twig to which it is attached.*

2 *Bring all the leaves back to the classroom.*

3 *Take one leaf and examine it carefully (Fig. 4.15).
Look for the following parts, and try to learn their
names and what they do:*

- ***petiole****, or stalk, that attaches the leaf to the plant.*
- ***lamina****, or leaf blade. This is usually thin and
  flattened, with a big surface area to collect light
  from the sun.*
- ***margin*** *– the outer edge of the lamina.*
- ***apex*** *– the tip of the leaf.*
- ***midrib****, or main central vein, leading in from the
  petiole. In the broad-leaved plants the midrib gives
  off branching veins.*
- ***veins****, branching to all parts of the leaf. The veins
  support the thin lamina. They contain xylem and
  phloem vessels, the thin tubes that transport water
  and dissolved food to and from the leaf.*
- ***cuticle****, on the upper surface, and made of wax.
  This makes the upper surface shinier than the
  lower surface. The cuticle prevents too much
  water from being lost by evaporation.*
- ***pores*** *(or* ***stomata****) on the lower surface. They
  allow the leaves to exchange gases and water
  vapour with the air. You would need a microscope
  to see these tiny holes.*

4 *Sort out the different* **leaf shapes***. See if you have
examples of each of the leaf designs shown by Fig.
4.16.*

5 *Now look at the* **leaf margins***. Again there are
different shapes. See if you have the two shown by
Fig. 4.17 in your collection.*

6 *Now look at the arrangement of the veins of your
leaves. Look for the two types of venation shown by
Fig. 4.18.*

7 *The total surface area of all the leaves on one tree is
enormous. You can estimate this area by working out*

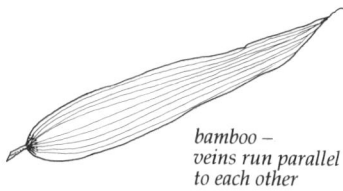

bamboo –
veins run parallel
to each other

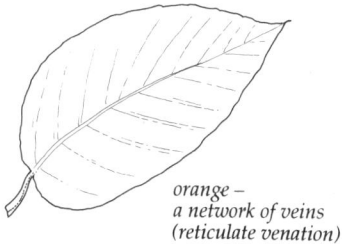

orange –
a network of veins
(reticulate venation)

Fig. 4.18 *Types of leaf venation.*

Fig. 4.19 *Finding the area of a leaf.*

the surface area of one leaf in square centimetres, and then multiplying this by the number of leaves.

To do this, place a leaf on a piece of graph paper and draw round the edges. Count all the complete squares inside the outline to find the surface area. Where the outline cuts across small squares, count up all these squares and divide by two to get an estimate of the area around the margin. Add this to the number of complete squares to get the total area of your leaf.

8 When you know the area of one leaf, you can work out the approximate total area of all the leaves on a tree. Go back to the tree.

(a) Count the number of leaves on one small branch.

(b) Count the number of similar branches on the tree.

(c) Multiply the number of leaves by the number of branches to find the total number of leaves.

(d) Then multiply the area of one leaf by the total number of leaves. This will give you the total area in square centimetres. To convert this into square metres, divide by 10,000.

You might be surprised to find how great an area of leaves the tree can expose to the sunlight.

9 You can add to your plant collection by pressing and mounting the leaves of different trees and shrubs. This is a useful exercise, because when you know the different leaf forms you will become better at recognizing crop plants and can more easily tell the difference between weeds and crop plants.

The functions of leaves

Leaves have the following functions:

1 *Photosynthesis.* This is the process by which green leaves make food in the presence of sunlight. This process will be discussed in more detail later.

2 *Respiration* . Plants obtain energy for growth by breaking down carbohydrates. This process is called respiration, and it takes place in all living parts of the plant, including the leaves.

3 *Transpiration.* This is the name for the loss of water vapour by diffusion through the pores, (or **stomata**), of the leaves.

4 *Storage.* The leaves of some plants are very thick and swollen, and used for storing food and water. Examples of plants with leaves like this are the aloe and the onion.

5 *Reproduction.* Some leaves produce small plants around their margins that drop off and take root. This is unusual, but it does happen. An example of a plant with leaves like this is called 'Wonder of the World.'

6 *Protection.* Sometimes leaves are used to protect parts of the plant. An example is the onion bulb, where dry, scale leaves on the outside protect the soft swollen storage leaves inside.

Leaves are the 'factories' of a plant. They are the places where the plant makes food. Note how the structure of a leaf – its thin, flattened shape, its good water supply and system of veins – is very well suited to the work it has to do.

Questions

1 Draw a fully labelled diagram of a broad leaved flowering plant.

2 (a) List four functions of roots.

(b) Name two plants with unusual root systems. Explain the special functions of their roots.

3 (a) List three functions of stems.

(b) Draw labelled diagrams of three different types of shoot system. Explain their functions.

4 (a) Name one plant which is propagated from suckers growing up from the root system.

(b) Name the crop propagated by this type of shoot which Captain Bligh of the Bounty brought from Tahiti in the South Seas to the West Indies, in 1793.

(c) What notorious event happened on that voyage?

(d) Why was this crop imported?

5 Explain why a leaf can be described as a 'food factory'.

5

LIVING PROCESSES OF PLANTS

When you have read this chapter you will be able to:

- *explain what is meant by photosynthesis.*
- *write the equation for photosynthesis.*
- *list the conditions required for photosynthesis to take place.*
- *describe experiments that show these conditions are necessary.*
- *describe how to test a leaf for the presence of starch.*
- *explain how to improve the photosynthetic efficiency of crops.*

Every living organism, whether plant or animal, carries out seven **living processes**. These are the things they must do to stay alive, and increase their numbers. There is no living organism that does not do these things, and *only* living organisms can do them. Here is a list of the seven processes:

Growth

Respiration

Irritability (response to surroundings)

Movement

Nutrition

Excretion

Reproduction

We recognize these processes because they are part of our own way of life. Can you say how human beings carry out each process?

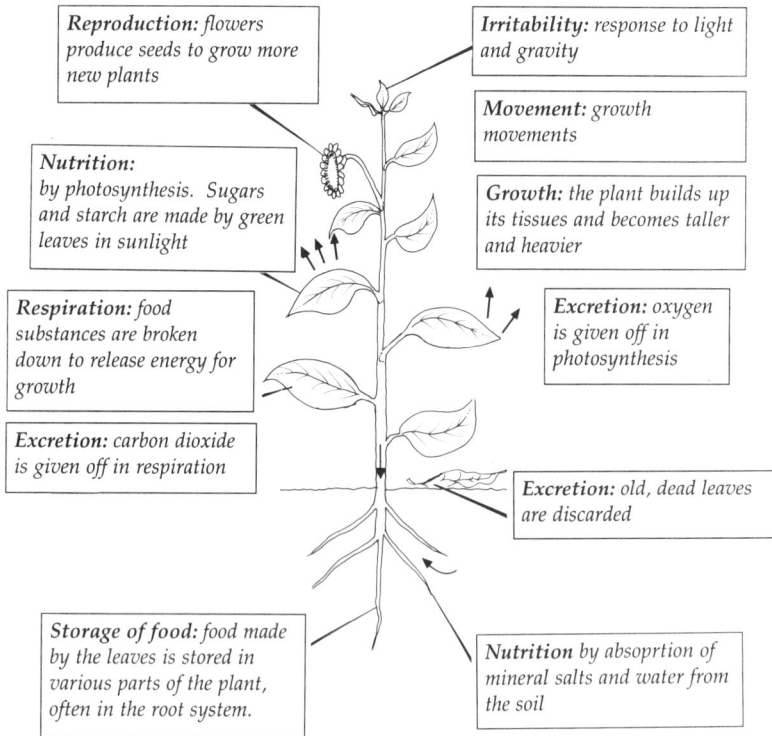

Reproduction: flowers produce seeds to grow more new plants

Irritability: response to light and gravity

Movement: growth movements

Nutrition: by photosynthesis. Sugars and starch are made by green leaves in sunlight

Growth: the plant builds up its tissues and becomes taller and heavier

Respiration: food substances are broken down to release energy for growth

Excretion: oxygen is given off in photosynthesis

Excretion: carbon dioxide is given off in respiration

Excretion: old, dead leaves are discarded

Storage of food: food made by the leaves is stored in various parts of the plant, often in the root system.

Nutrition by absoprtion of mineral salts and water from the soil

Fig. 5.1 *Living processes of plants.*

As you know already, the way of life of plants is very different from ours. They do everything so slowly! They make no sound, they stay rooted to the ground, and do not seem to get excited whatever happens to them. Even so a plant, in its own way, carries out exactly the same vital processes as an animal or a human being. Green plants can also carry out **photosynthesis**, which animals cannot.

Living processes of plants

1 **G**rowth   – getting bigger and heavier.

2 **R**espiration   – producing energy for growth and maintenance, using up food reserves.

3 **I**rritability   – plants respond to their surroundings by growing in certain directions.

4 **M**ovement   – growth movements, in the case of plants.

5 **N**utrition   – through the roots, and by photosynthesis in the leaves.

6 **E**xcretion   – getting rid of waste products.

7 **R**eproduction   – making new plants from parent materials.

Notice that the initial letters of the living processes form the acronym **GRIMNER**. There is really no such word, but if you can remember **GRIMNER**, it will then help you to remember the names of all the processes.

It takes patience to discover how plants live their lives. Sometimes experiments lasting for several days are needed to find out what plants are doing. But it is important to understand how plants live, because this will enable us to meet their needs, and help them to produce good crops.

Nutrition is one of the basic vital processes, and in plants this is achieved partly by photosynthesis. This may not *seem* like nutrition at first sight, but it is one of the ways in which plants feed.

Photosynthesis

This is what green plants do in sunlight. Photosynthesis is a long word, but not too hard to understand. 'Photos' is a Greek word meaning 'light', and 'synthesis' means 'building up'. The plant is 'building something up with the help of light'. The substance it is building up is **sugar**. This is why photosynthesis is actually a form of feeding for plants.

The chemical reactions of photosynthesis can be summarized briefly by an equation:

(raw materials)   (energy source)

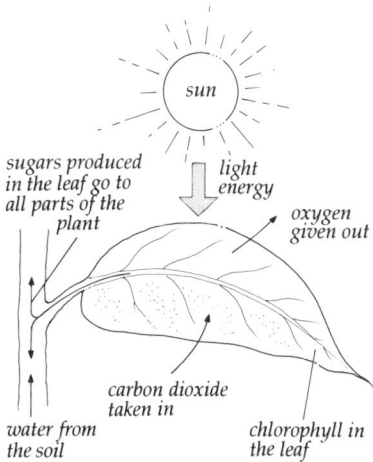

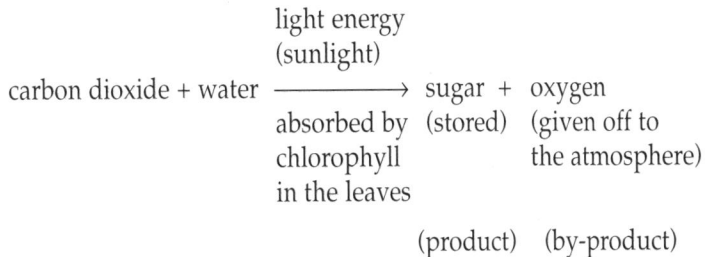

Fig. 5.2 A green plant carries out photosynthesis.

light energy
(sunlight)

carbon dioxide + water ⟶ sugar + oxygen
absorbed by (stored) (given off to
chlorophyll   the atmosphere)
in the leaves

(product) (by-product)

This equation tells you which substances are reacting together, and what the products are at the end. The **raw materials**, at the start of the reaction, are carbon dioxide and water. The **products**, at the end, are sugar and oxygen. Read the equation slowly, from left to right, and try to follow what is going on, one step at a time:

1 **Carbon dioxide** gas is taken in from the air through the pores of the leaves.

2 **Water** is taken in through the roots and passed up to the leaves. So carbon dioxide and water are both present in the leaf.

3   When the sun shines on the leaf, the water and carbon dioxide combine together to make sugar. This happens only in the presence of **chlorophyll**, (the green substance in plant leaves, which absorbs the light energy of the sun). The leaf is the reaction site, or 'factory', where the reaction of photosynthesis takes place.

4   **Sunlight** is needed to make this reaction happen. Light supplies energy to drive the reaction. It cannot work without light.

5   **Oxygen** gas is given off in photosynthesis. It is released into the air through the pores of the leaves. This is a good thing for us! We need oxygen for respiration, and so do all animals, and the plants themselves.

6   Sugar is moved about inside the plant through tiny tubes, called **phloem vessels**, which you have heard about already. The plant can store food in the form of sugar, starch or oil. It may store this food in almost any part of its body – in its leaves, stems, roots, flowers, fruits or seeds. Can you name any plants you know which store food in these various parts?

Photosynthesis is the plant's way of making food for itself. First it makes sugar. Then it goes on to make other substances from this sugar – things like carbohydrates, fats, oils, proteins, vitamins – in fact all the foods *we* need to eat. We get a lot of our food from plants. Of course, the plants are not 'thinking' about us! They are just growing, and building up their own bodies.

Why photosynthesis is important

All life depends on photosynthesis This means that photosynthesis is probably the most important chemical process taking place on earth. Green plants are the *only* living creatures that can make the foods needed by animals and ourselves. Not only do green plants give us food, they supply the oxygen in the air we breathe. Without them we would not be here at all, would we?

Experiments on photosynthesis

Remember that an experiment is just a way of finding something out. When you do an experiment, you are

growing in the light

growing in the dark

Fig. 5.3 *Growing seedlings in the light and in the dark.*

growing in full daylight

growing on a windowsill

Fig. 5.4 *The effect of the direction of light on plant growth.*

saying, 'Let's try this, and see what happens.' You may be able to do some of these experiments in class, or your teacher may demonstrate them for you. Some of the experiments may be carried out in your Integrated Science class, or you may do them as group assignments, and share the written results and conclusions.

The first two experiments show that plants need light to grow properly. The rest show how photosynthesis works.

PRACTICAL WORK
Experiment 1 To show that a plant needs light to make chlorophyll

1 *Plant two bean seeds in separate containers, (plant pots or milk cartons), containing soil.*

2 *Water them well.*

3 *Put one in a dark cupboard where it has no light. Leave the other in a place where it has plenty of light.*

4 *Watch the two seedling plants that grow over the next week.*

5 *Draw a diagram of the experiment.*

6 *Record your observations.*

7 *What are your conclusions?*

Experiment 2 To show that a plant shoot grows towards light

1 *Take two young bean plants growing in separate containers.*

2 *Put one on a windowsill, where it will get light from just one side.*

3 *Put the other plant out in the garden, where it will get light from all sides.*

4 *Keep the plants watered, and watch what happens to them after a few days.*

5 *Draw a diagram of the experiment.*

6 *Write down what you saw, and what you found out.*

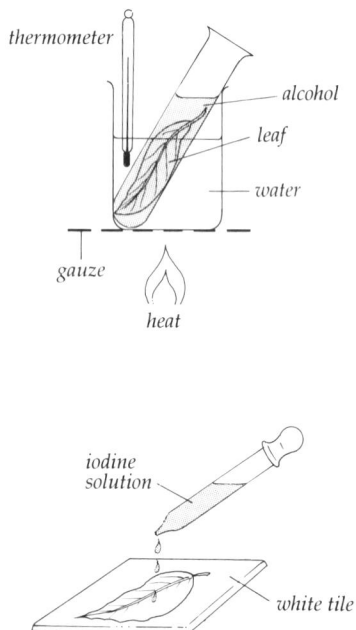

Fig. 5.5 Testing a leaf for starch.

How to tell if a plant has been photosynthesising. As soon as a leaf has made some sugar, it converts some of it into starch. You know what starch is. It is the white powdery substance found in flour and in potatoes. Starch does not dissolve in water, and it cannot move about in solution. The plant uses it for storage for this reason. It stays where it is put.

Experiment 3 To test a leaf for the presence of starch

Starch can be detected by a simple test. Take some flour, and put a drop of iodine solution on to it. It goes a dark, blue/black colour, like ink. If you can produce this colour in a leaf, with iodine solution, you know that the leaf has been carrying out photosynthesis and making starch.

1 Take a leaf from a plant that has been in bright sunlight for a few hours. First kill the leaf by dipping it in boiling water. This stops any chemical reactions that may be going on inside, and it softens the leaf.

2 Next remove the chlorophyll from the leaf using hot alcohol. Put the leaf in a big test-tube, and heat it gently in a water bath as shown. The boiling point of alcohol is 72 degrees Celsius. If you use a thermometer, you can warm the alcohol until it just boils. Do not boil it too hard, because the vapour that comes off can catch fire.

3 After about ten minutes, the chlorophyll will have dissolved out of the leaf, and the alcohol will be coloured green.

4 Now take the leaf out of the alcohol, using forceps. At this stage it will be light brown and very brittle. Dip it in hot water again to soften it, and spread it out on a white tile or plate.

5 Add a few drops of iodine solution, and give it about five minutes to soak in. The leaf will go blue/black in colour, showing that starch has been made by photosynthesis.

6 A *control experiment* can be carried out at the same time. Take a leaf from a plant that has been kept in a dark cupboard for 24 hours. Test it in the same way, and you will find that it contains no starch. This is

Fig. 5.6 *A plant with variegated leaves.*

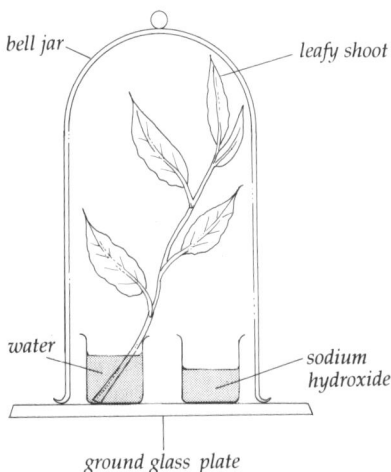

Fig. 5.7 *Carbon dioxide is needed for photosynthesis.*

because the plant has used up the starch in its own respiration.

A control is always exactly the same as the original experiment, except for the one *factor being tested. It proves that this is the factor that makes the difference.*

7 The test for starch is done after each of the following experiments, to check if photosynthesis has taken place.

8 Draw a diagram of the experiment.

9 Write down what you saw, and what you found out.

Experiment 4 To show that chlorophyll is necessary for photosynthesis

Remember that chlorophyll is the green substance in the leaves of plants.

1 For this experiment you need a plant with variegated leaves. This means a plant whose leaves are partly green and partly yellow or white, e.g. variegated hibiscus (Hibiscus acalypta) or croton (Codiaeum variegatum), or variegated Spanish thyme (Thymus vulgaris variegatum).

2 After it has been in the light for a few hours, take a leaf and test it for starch, as shown in the previous experiment.

3 Only the parts that were green will have photosynthesised. The green parts contained chlorophyll and the yellow parts did not. Therefore, chlorophyll is necessary for photosynthesis to take place.

4 Draw a diagram of the experiment.

5 Write down what you saw, and what you found out.

Experiment 5 To show that carbon dioxide is necessary for photosynthesis

1 Place a cut plant shoot in water under a bell jar, exposed to light. Put a beaker containing strong alkali, such as sodium hydroxide, under the jar beside

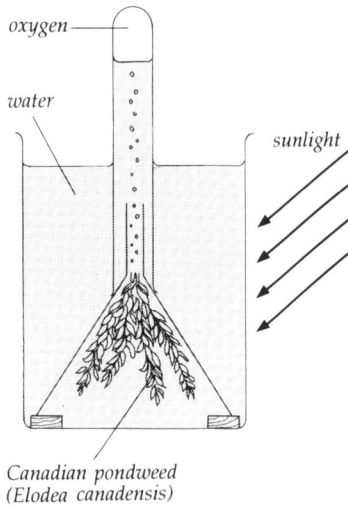

Fig. 5.8 Oxygen is given off during photosynthesis.

Fig. 5.9 Testing for oxygen.

the plant. This will remove all the carbon dioxide from the air inside the bell jar. *The bell jar fits on to a ground glass plate smeared with vaseline. This makes it airtight.*

2 *After about six hours, a leaf from the plant is tested for starch. No starch is found, showing that no photosynthesis has taken place. Carbon dioxide is therefore necessary for photosynthesis.*

3 *A control experiment can be set up, without the alkali. In the control, photosynthesis does occur and starch is found.*

4 *Draw a diagram of the experiment.*

5 *Write down what you saw, and what you found out.*

Experiment 6 To show that oxygen is given off in photosynthesis

The problem here is that you have to catch some oxygen coming off a plant. Then you have to test the gas to see if it is oxygen. This would be hard to do with a plant growing in the open air. The way round the problem is to use a water plant. A healthy-looking pondweed will do.

1 *Put the plant under an inverted funnel in a large beaker of water, as shown in Fig. 5.8. Fill a test tube with water and invert it over the funnel. Any gas coming from the plant will collect in the test tube.*

This is not as easy as it looks! One way of doing it is to put the apparatus together in a sink full of water. Then, when it is ready, you can lift it out, and set it up on the bench.

2 *Add a pinch of sodium bicarbonate to the water to make sure that it contains enough carbon dioxide in solution. The plant needs the carbon dioxide for photosynthesis.*

3 *Put the apparatus in sunlight. You should soon see bubbles of gas coming off it, and rising to the top of the test-tube. If photosynthesis has been taking place, this gas should be mostly oxygen.*

4 *Carefully remove the test-tube, with the gas still in it, and put your thumb over the end. Remove your* *thumb and put a glowing splint of wood into the tube. If it glows more brightly and begins to burn, the gas is oxygen.*

A result like this shows that oxygen was given off during photosynthesis.

5 *Draw a diagram of the experiment.*

6 *Write down what you saw, and what you found out.*

Experiment 7 To show that light is necessary for photosynthesis

1 *Place a potted plant in the dark, (e.g. in a cupboard), for 24 hours. This **destarches** it. The plant uses up any starch in its leaves as it respires.*

2 *Bring the plant out and fit a 'light screen', made of cardboard or aluminium foil, on to one of its leaves. It is a good idea to cut a recognizable shape, (e.g. a triangle or a square), in the light screen before putting it on to the leaf.*

3 *Leave the plant in bright light for six hours. Then take off the leaf, remove the light screen, and test the leaf for the presence of starch. Only the parts that received light go blue/black, showing that they have photosynthesised.*

4 *Draw a diagram of the experiment.*

5 *Write down what you saw, and what you found out.*

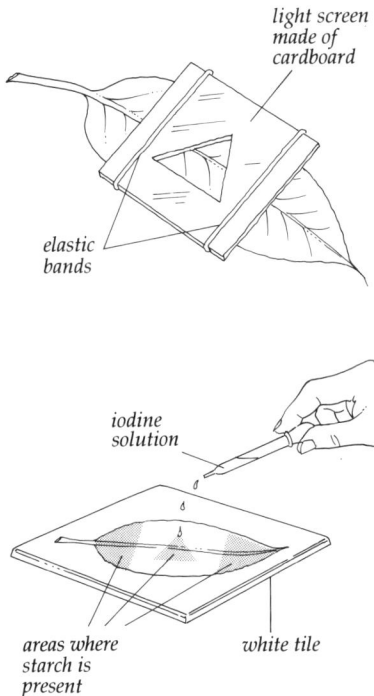

light screen made of cardboard

elastic bands

iodine solution

areas where starch is present

white tile

Fig. 5.10 Light is needed for photosynthesis.

Agricultural implications of photosynthesis

Since photosynthesis is so important for the production of food by plants, which man and animals also need, plants need to have the best conditions to carry out photosynthesis.

1 The plant must be able to display all its green leaves to the sunlight.

2 It must have water and nutrients.

To ensure this:

1 Plants must be spaced properly so that there is no self shading.

2 Weeds should be removed to prevent them from robbing the plant of light, water and nutrients.

3 Plants must have an adequate supply of water in a well aerated soil so that water and nutrient uptake is made more easy.

4 Plants should be grown in places where they can get at least six hours of direct sunlight every day.

5 Leaves should be kept free of diseases such as sooty mould fungus which can reduce the amount of chlorophyll in them, and so reduce the absorption of light energy.

6 Trees should be pruned and shaped to achieve these objectives.

Ensuring or enabling photosynthesis to be carried out by the plant is described as improving **photosynthetic efficiency**.

Can you answer these questions:

1 Plants and people

Why do people need plants? Can you name three things we get from plants?

Why do plants need people? Can you think of three things that plants get from people? (This is a harder question!)

Do you think we need plants more than they need us?

2 The foods we eat

Can you name the nutrients present in the foods we eat?

Where do they come from?

How are they made?

3 Why we need green plants

Would there be any life on earth without the green plants?

Why not?

If you can answer these questions, you have learned some very important facts about the relationship between people and plants.

Questions

1 (a) Define 'photosynthesis'.

   (b) Write down the chemical equation for photosynthesis.

2 (a) Name three conditions needed for photosynthesis to take place.

   (b) With the help of a diagram, describe an experiment to show that any one of these conditions is necessary.

3 (a) Describe, with the help of diagrams, how to test a leaf for the presence of starch.

   (b) How can you 'destarch' the leaves of a living plant?

4 (a) List briefly the seven 'living processes' of organisms.

   (b) How can plants be said to 'move'?

5 (a) Describe, with the help of a diagram, an experiment to show that oxygen is given off in photosynthesis.

   (b) How can you tell that the gas given off is oxygen?

6

FLOWERS AND SEXUAL REPRODUCTION

When you have read this chapter you will be able to: .

- *draw labelled diagrams of:*
    - *an insect-pollinated flower*
    - *a wind-pollinated flower.*
- *explain what each part of the flower does.*
- *say what is meant by pollination, and how it takes place.*
- *say what is meant by fertilization, and how it takes place.*
- *describe the changes taking place in a flower after fertilization, and the formation of fruits and seeds.*
- *explain the functions of fruits.*
- *show how plants disperse their seeds.*

We like flowers because they have beautiful shapes, colours and scents, but plants do not produce flowers just to keep *us* happy! Flowers are the reproductive organs of plants. They produce seeds, and the seeds start the next generation of plants. The production of new plants is called **propagation**.

Flowers come in all shapes and sizes, but they all have the same basic parts. Most flowers have both male and female organs arranged inside their petals. They produce **pollen** (male cells) in their **anthers**, and **ovules** (female cells) in their **ovaries**.

Pollen is transferred from one flower to another by insects or by the wind. We call this transfer of pollen **pollination**. When the male cells join with the female cells, we call this **fertilization**.

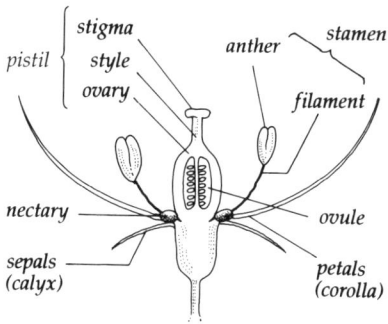

Fig. 6.1 *Vertical section through a flower.*

The structure of a flower

Note what each part does:

- **Sepals** enclose the flower bud before it opens. All the sepals together make up the **calyx**.
- **Petals** are brightly coloured to attract insects and these flowers often have a sweet scent for the same purpose. All the petals in each flower together make up the **corolla**.
- **Anthers** and **filaments** are the **male organs** that produce pollen – a yellow dust-like substance. Each pollen grain contains a male cell. The filaments are the stalks that hold up the anthers. An anther and filament together are called a **stamen**.
- **Ovaries**, **styles** and **stigmas** are the **female organs**. The stigma is sticky, for collecting pollen grains from other flowers. The style leads down to the ovary. Inside the ovary are the ovules. Each ovule contains a female cell that can be fertilized by pollen. A stigma, style and ovary together are known as a **carpel**.
- **Nectaries** make a sweet, sugary liquid called **nectar** that insects like. This attracts insects to the flower, and makes them unsuspecting agents of pollination.

Plants cannot move about, so they rely on insects or the wind to transfer their pollen from one plant to another. **Insect-pollinated flowers** are very different to look at from **wind-pollinated** ones. More about that later on pages 46 and 47.

PRACTICAL WORK
Examining the parts of a flower

1 *Go out into the school compound, and see if you can watch insects visiting flowers. Notice how they move busily from flower to flower looking for nectar. Can you see pollen sticking to the hairs of their bodies?*

2 *Collect some brightly coloured insect-pollinated flowers. Take just one good flower from as many different kinds of plants as you can.*

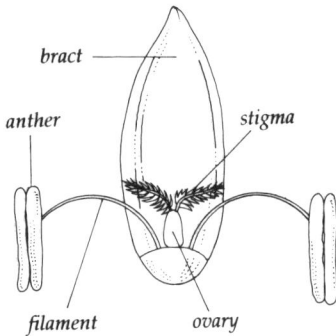

Fig. 6.2 A wind-pollinated grass flower.

3 Collect some wind-pollinated flowers. Examples of these are the grasses. See if you can find some ripe ones, with anthers and stigmas sticking out. Take one of each kind.

4 Back in the classroom, look inside the insect-pollinated flowers. Can you find the parts shown in Fig. 6.1? All flowers have these parts, though they may be shaped differently.

5 Examine the wind-pollinated flowers with a hand lens. Notice that they have no coloured petals, and no sweet scent. Look for the large anthers on flexible filaments, and for the feathery stigmas, shown in Fig. 6.2.

6 Draw and label a grass flower, to help you remember the parts.

7 If you have a microscope, make mounts of different kinds of pollen. Examine and compare them. Note that grass pollens, carried by the wind, have smaller, lighter grains than pollens carried by insects. Look also at slides of ovaries containing ovules.

Pollination

Insect-pollinated flowers

These try hard to attract the insects to visit them. They have brightly coloured petals, a pleasant scent, and sweet nectar. The insects are not thinking about helping the flowers when they visit them! They come for the nectar.

Fig. 6.3 A flower being pollinated by a bee.

When a bee crawls over a flower, pollen sticks to the hairs of its body. When it goes to a flower that has a ripe stigma, some pollen will be left behind on its sticky surface. Look at the picture of a bee visiting a flower (Fig. 6.3). Can you see how it picks up pollen and transfers it to another flower? This transfer of pollen from the anther to the stigma is called pollination. The relationship between plants and insects works very well for them both.

Transfer of pollen to the stigma of the *same* flower is called **self-pollination**. Transfer of pollen to the stigma of *another* flower is called **cross-pollination**.

Wind-pollinated flowers

These do not try to attract insects. Examples of plants with this type of flower are grasses and cereal crops. They have no pretty petals, scent or nectar. Instead, they grow big anthers that hang outside the flowers on flexible stalks or filaments. The anthers shed a lot of small, light pollen grains that blow about on the slightest breeze. The stigmas of these flowers stick out from the flower, and have a branching, feathery structure. This helps them to 'comb' the wind for pollen, which sticks to the surface of the stigmas.

You might think that flowers would pollinate themselves, as they have both male and female parts, but these parts do not usually ripen at the same time. This means that most flowers are cross-pollinated. The pollen from one flower fertilizes the ovules of another flower. Some plants are self-pollinated, but most are not.

Fertilization

When a pollen grain arrives on the stigma, it grows a long, thin **pollen tube** down through the style. This tube enters one of the ovules in the ovary. Then the **male nucleus** in the pollen grain passes down the tube and fuses (joins) with the **female nucleus** of an ovule. This is called fertilization.

This is a sexual process, like that which occurs in animals. The pollen is like the sperm and the ovule is like the egg. A seed formed by this process will have *two parents*. The plant that grows from the seed will have features of both of them.

Try not to confuse pollination with fertilization! It is easy to mix them up.

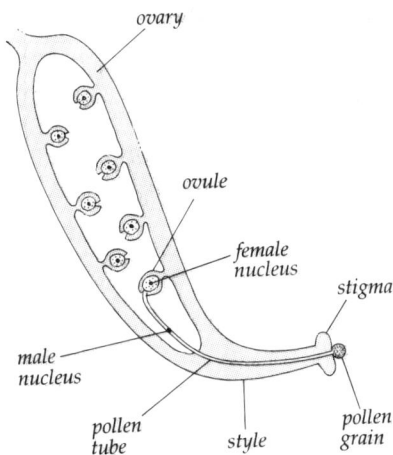

Fig. 6.4 *A pollen grain fertilizing an ovule.*

> **Pollination** means the transfer of pollen from the anthers to the stigma. Pollination is completed before fertilization begins.

> **Fertilization** means the fusing of the male pollen grain nucleus with the female ovule nucleus. When these two nuclei fuse they form a **zygote** which then develops into an **embryo**.

Fruit and seed formation

After fertilization, big changes take place:

- The petals of the flower die and fall off. They are no longer needed to attract insects.
- The ovary grows much bigger and becomes the **fruit**. There are many different shapes and types of fruit, but they are all formed in this way.
- The fertilized ovules in the ovary grow to become the **seeds** of the fruit. Each seed has a tiny **embryo** plant inside it that may some day grow into a new plant.

PRACTICAL WORK

Looking at fruits and seeds

1 *Go back after two or three weeks to the plants in the school compound where you collected your flowers.*

2 *See what has happened to the flowers. Look for fruits forming from the ovaries, and seeds forming inside the fruits.*

3 *Collect some of these fruits.*

4 *Back in the classroom, open the fruits to see the seeds. Draw them and label the parts.*

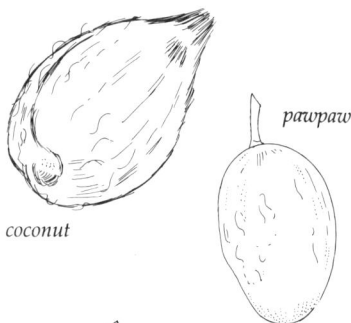

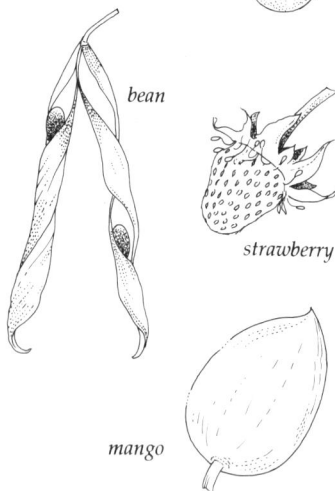

Types of fruits

Fruits are **classified** into groups by their appearance and texture. Some are dry, some are juicy. Some burst open and shed their seeds, (the **dehiscent** fruits). Some keep their seeds inside them, (the **indehiscent** fruits). These are the main groups, with an example of each:

Group	Example
Dry fruits	coconut
Juicy fruits	mango
Dehiscent fruits	bean
Indehiscent fruits	paw paw
False fruit	cashew

coconut

pawpaw

bean

strawberry

mango

Fig. 6.5 *Types of fruits – note the differences in appearance.*

Mango: the fruit is eaten and the seed is dropped

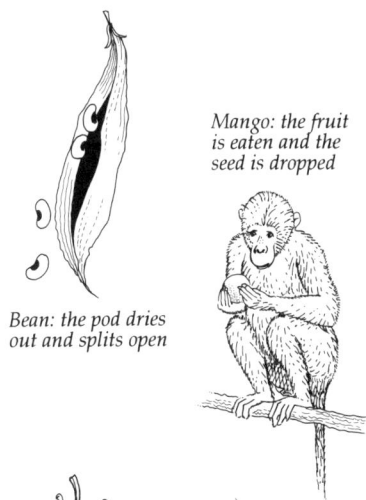

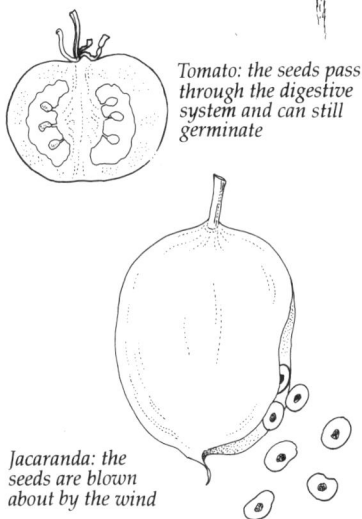

Bean: the pod dries out and splits open

Tomato: the seeds pass through the digestive system and can still germinate

Jacaranda: the seeds are blown about by the wind

Fig. 6.6 *How plants spread their seeds about.*

The cashew is sometimes called a **false fruit**, because it does not develop from the swelling of the ovary. Instead, the **receptacle** swells up to form this false fruit. The real fruit of a cashew has a coat which is thick, hard and shiny on the outside. It looks like a seed, but it is really a small fruit containing a single dicotyledonous seed which is the part we eat.

Functions of fruits

1 Fruits store food made by the plant, but the plant itself does not use this food.

2 Plants rely on animals to feed on the fruits, and to carry the seeds some distance away from the parent plants. Some of these seeds then fall to the ground and germinate to produce new plants. The fruit is therefore a seed dispersal mechanism.

Seed dispersal

Seed dispersal means 'spreading seeds about'. Plants have many ways of dispersing their seeds. If they did not do this, their seeds would fall close to the parent plant. Then the seedlings would have to compete with each other for space, light and water. Fig. 6.6 shows some of the clever ways by which plants disperse their seeds.

The Bee

Busy bee
Don't bother me.
Being busy
Makes me dusy.
All the bee does
Is boes ... boes ... boes.

Frank Collymore

Questions

1 (a) Draw a fully labelled diagram of a wind-pollinated flower seen in section.

(b) What is the function of:
   - the sepals?
   - the anthers?
   - the ovary?

2 (a) Name three ways in which insect pollinated flowers attract insects.

(b) Describe how pollen is transferred by a bee from the anthers of one flower to the stigma of another.

3 (a) Draw a fully labelled diagram of a wind pollinated flower.

(b) Describe three features that make it easy for the wind to transfer pollen from the anthers of one flower to the stigma of another.

4 (a) What is the difference between 'pollination' and 'fertilization'?

(b) Explain, with the help of a diagram, how a pollen grain on the stigma is able to fertilize an ovule in the ovary of the flower.

5 (a) Describe three changes that take place in the flower of a named plant after fertilization.

(b) How are the seeds of this plant dispersed?

7

GROUPS OF CROPS

When you have read this chapter you will be able to:

- *say what is meant by classification.*
- *describe the **binomial system** of naming plants and animals.*
- *group some crop plants:*
    - *into their families*
    - *according to their food value*
    - *according to their end use.*

Classification

Classification means grouping similar things together and arranging them in a logical way.

PRACTICAL WORK

Groups of crops

1 *Make some cards measuring 1 cm × 5 cm. Write the words from the list given below on these cards.*

tomato	*carrot*	*black eye pea*
corn	*sweet pepper*	*banana*
salad beans	*sweet potato*	*coconut*
yam	*orange*	*soya bean*
melongene	*rice*	*bodi*
sugar cane	*beet*	*eddoes*
peanut	*nutmeg*	

2 *Arrange the cards into groups, putting similar plants together. As you find these groups, write the list in your notebook.*

Fig. 7.1 Solanum melongena. What is its common name?

A short history of classification

Aristotle the Greek philosopher appears to have been the first person to try to classify (or group) plants by comparing their characteristics.

Carolus Linnaeus, the Swedish botanist, is regarded as the founder of modern classification, or **taxonomy**. He gave each plant and animal *two* Latin names. One was its **generic** (or **family**) name, and the other was its **specific** name, to define its place within its family. This method is now called the 'binomial system of nomenclature' – which just means that every organism has two names. The system we use today for naming organisms is based on the rules of Linnaeus.

An example may help to show how this naming system works. The egg plant is called *Solanum melongena*.

The first name, *Solanum*, shows the **genus**, or family group, to which the plant belongs. This group includes tomatoes, potatoes and many other plants. They are similar to each other in structure, and are descended from common ancestors. The second name, *melongena*, belongs *only* to this one plant. This is the specific name.

Why classify?

Why should we bother to arrange crops in groups like this? The answer is that scientists do need some common means of agreed identification so that they can communicate with each other.

The egg plant has a lot of common names – aubergine, bolounger, bhaigan, egg plant, melongene. This can be confusing, because the common name varies from place to place, and from one language to another, but a scientist in any part of the world will recognize the crop by its botanical name of *Solanum melongena*.

You may still wonder what use classification is to farmers and gardeners. The first thing to realize is that there are different 'criteria' (reasons) for grouping crops. For example, farmers may decide to group them according to their food value, or according to their families, or according to the use to which the crop is put.

Crops cannot be put into *one* group which satisfies *all* criteria, because they are similar in some respects and dissimilar in others. A crop may be grouped or classified into more than one group.

Two ways of classifying crops

1 Food value of the crop

This way is based on the major food supplied by the crop. Crops contain all major nutrients in varying proportions. However, there is usually one nutrient which occurs in a greater quantity. Here are some examples to remind you of the major nutrients, and some of the crops that supply them:

Carbohydrates – 'Go food'	Give energy to people and animals. Supplied by cereal grains, potatoes and sugar cane.
Proteins – 'Grow food'	Responsible for the growth and repair of tissues. Supplied by peas, beans and other legumes.
Fats and oils – 'Go food'	Give heat and energy. Supplied by groundnuts, sunflower seeds, and many other seeds.
Vitamins and minerals – 'Glow food'	Give healthy growth. Supplied by fresh vegetables.

Fig. 7.2 Carbohydrates for energy.

Fig. 7.3 Proteins for growth.

Fig. 7.4 Vitamins and minerals for health.

Fig. 7.5

narrow leaves with parallel veins

a) A maize plant is a monocotyledon.

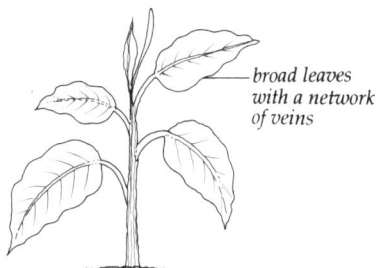

broad leaves with a network of veins

b) A bean is a dicotyledon.

High fibre foods	Contain a high percentage of cellulose.
– 'Go food'	For **polygastric** animals, i.e. those with compound stomachs, fibre is their main source of energy.
– 'Glow food'	For **monogastric** animals, those with simple stomachs, fibre provides bulk and also keeps the animal's alimentary system healthy. Supplied by all plants in the form of **cellulose** which makes up their cell walls.

2 Plant families

In classifying plants which make up most of our crops two sub-classes can be recognized:

(i) **monocotyledons** plants with narrow leaves and parallel veins, e.g. the cereal crops.

(ii) **dicotyledons** plants with broad leaves and veins arranged in a network formation, e.g. most other crops.

These main sub-classes are further divided into families of plants that are more closely related. Fig. 7.6 shows some of the main families to which crop plants belong.

Family	Some family members
Solanaceae	tomato, melongene, sweet pepper, hot pepper
Cruciferae	cabbage, patchoi, mustard, broccoli
Gramineae	corn, rice, elephant grass, pangola grass, bamboo, coconut etc. The grass family is the largest plant family.

Fig. 7.6 Crop plants and their families.

Members of plant families usually require the same type of cultivation practices, and are attacked by similar diseases and pests. This influences the farmer in planning crop rotations.

Exercise

The table below shows two methods of classification –
according to:

1 The end use of the crop (in the second column), and
2 The botanical names (in the third column).

Complete the table by filling in the blank spaces.

Crop	1 End use of crop	2 Botanical name	Common name
Vegetable (a) Fruit			melongene
(b) Leafy		*Lactuca sativa*	
(c) Root			carrot
Legumes and pulses		*Phaseolus vulgaris*	string bean
Food crops (a) Root tuber		*Manihot esculenta*	
(b) Rhizome		*Zingiber*	
(c) Stem tuber		*Diascorea alata*	
(d) Fruit		*Musa* species	plantain
Grasses (a) Pasture	Food for animals		pangola grass
(b) Fodder		*Paspalum faciculatum*	
(c) Cereal		*Oryza sativa*	
(d) Stem crop		*Saccharum officinarum*	
Fibre		*Gossypium* sp.	
Fruit crops	fresh fruit, processing		long mango
Nuts		*Arachis hypogaea*	
Oils and fats		*Glycine max*	
Medicinal plants			aloe
Forest trees			teak
Spices and condiments			hot pepper
Ornamental crops		*Hibiscus rosa-sinensis*	

Questions

1 (a) What is meant by 'classification'?

   (b) Name   (i) four root vegetables
               (ii) four fruit vegetables
               (iii) four cereal grains.

   (c) Why is it helpful to classify plants?

2 (a) Why do plants need Latin names, and why are their common names not always sufficient?

   (b) Name three crop plants that belong to each of the following families:
   (i) Solanaceae (ii) Cruciferae (iii) Gramminae.

3 (a) List the letters A to N in the margin of your book. Beside each letter write what you think is missing from the labelled spaces in the following table.

Crop	1 End use of crop	2 Botanical name	Common name
Vegetable (a) Fruit	A	B	melongene
(b) Leafy	C	*Lactuca sativa*	D
(c) Root	E	F	carrot
Legumes and pulses	G	*Phaseolus vulgaris*	string bean
Food crops (a) Root tuber	H	*Manihot esculenta*	I
(b) Rhizome	J	*Zingiber*	K
(c) Stem tuber	L	*Diascorea alata*	M
(d) Fruit	N	*Musa* species	plantain

4 (a) How can you tell that a plant is a monocotyledon?

   (b) How can you tell that a plant is a dicotyledon?

   (c) Give an example of a crop plant belonging to each of these groups.

5 Crops can be classified by their food value.

   (a) Name a crop that supplies mainly *carbohydrate*. Explain how animals make use of this food substance.

   (b) Name a crop that supplies mainly *protein*. Explain how animals make use of this food substance.

8

GROWING PLANTS FROM SEEDS

When you have read this chapter you will be able to:

- *draw a section through a bean seed, label the parts, and explain what each part does.*
- *explain what is meant by a dicotyledon.*
- *draw a section through a maize seed, label the parts, and explain what each part does.*
- *explain what is meant by a monocotyledon.*
- *describe experiments to show the conditions necessary for germination, (water, air and a suitable temperature).*
- *explain why these conditions are necessary.*
- *show how a seedbed provides good conditions for germination.*
- *say what is meant by the viability of seed.*
- *describe how to find the germination percentage of a sample of seed.*
- *explain why variation is an advantage to plants.*

PRACTICAL WORK
Examining seeds

1 *Soak some bean seeds and some maize seeds in water overnight. The next day they will be soft and easy to cut open.*

2 *Take a bean seed. Use a sharp knife to remove the **testa** or outer coat. Inside you will see that the seed is divided into two halves. These halves are the **cotyledons**, where food is stored to provide energy for growth. The bean belongs to a group of plants called the **dicotyledons**, (meaning plants with two*

Fig. 8.1

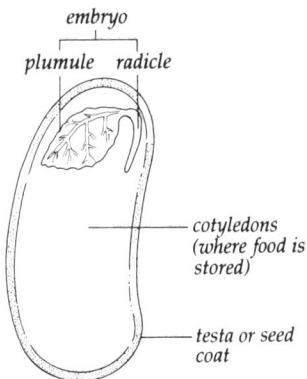

a) A bean seed cut in half.

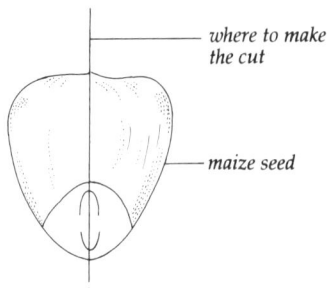

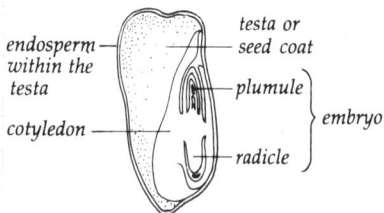

b) A maize seed cut in half and where to make the cut.

cotyledons). All the broad-leaved plants belong to this group. These plants also have **reticulated** leaf veins (i.e. veins that form a 'network' pattern) and tap root systems.

3 Separate the two cotyledons. Between them is what looks like a very small plant, with a tiny root and tiny leaves. This is the **embryo**. It has a **plumule** that will become the shoot, and a **radicle** that will become the root.

4 When a seed germinates, the embryo begins to grow. At first it uses the store of food in the cotyledons. When its shoot comes above the ground, it turns green and begins to photosynthesise.

5 Make a drawing of your bean seed cut in half, and label it with the names of the parts.

6 Now take a maize seed. Look for the position of the embryo. You can see it from the outside on the slightly raised portion on the white part of the seed. Cut the seed exactly in half with a sharp knife, and look at the embryo. Find the plumule and the radicle.

7 Between the embryo and the endosperm there is a shield-shaped structure. This is the cotyledon, and the maize seed has just one. If you compare its size to that of the bean seed you will realize that it does not store much food. The maize seed has an extra food store called the **endosperm**. The cotyledon's function is to absorb food for the embryo from the endosperm when the seed begins to grow. If you put some iodine solution on the cut surface of the seed, the endosperm goes blue/black in colour, showing the presence of starch. Maize belongs to a group of plants called the **monocotyledons**, meaning plants with just one cotyledon. Plants with narrow leaves and parallel veins (e.g. all the grasses) belong to this group. They also have fibrous roots.

8 Make a drawing of your maize seed cut in half, and label it with the names of the parts.

Experiments on germination

You can do experiments to find out how well plants germinate under different conditions.

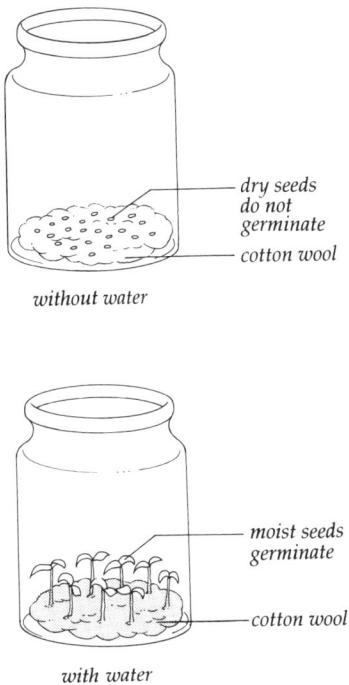

Fig. 8.2 *Germinating seeds need water.*

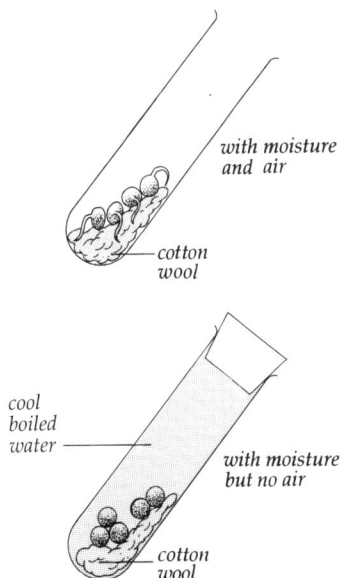

Fig. 8.3 *Germinating seeds need air.*

PRACTICAL WORK

Experiment 1 To show that water is necessary for germination

1 *Take two jars. Put some cotton wool in the bottom of each, and sprinkle on some pea seeds. Pea seeds are suitable because they germinate quickly, but other seeds could be used instead.*

2 *Add a little water to one of the jars to wet the cotton wool, and keep it moist for the next few days.*

3 *Leave the other jar without any water, so that the seeds stay dry.*

4 *Watch the jars for the next few days while germination takes place.*

5 *Draw a diagram of the experiment.*

6 *Write down what you saw, and what you found out.*

Experiment 2 To show that air is necessary for germination

1 *Take two large test-tubes. In the first one put moist cotton wool and some pea seeds. These seeds have both air and water.*

2 *In the second one put some cotton wool and pea seeds as before, but fill up the tube with water that has been boiled and allowed to cool. Put a cork in the test-tube to keep out the air. Now these seeds have plenty of water but no air.*

3 *Note that water usually has air dissolved in it. When you prepare the boiled water for the experiment, heat it in a glass beaker, and watch it as it heats up. Air bubbles will start to form on the sides of the beaker and escape to the surface. Boiling the water removes all the air dissolved in it. Make sure that the water is cooled to room temperature before pouring it on to the seeds.*

4 *Draw a diagram of the experiment.*

5 *Write down what you saw, and what you found out.*

6 *What gas from the air is needed for the respiration of germinating seeds?*

thermometer

in a warm,
shady place
(25°C)

on a hot
windowsill
(35°C)

inside a
refrigerator
(5°C)

Fig. 8.4 Germinating seeds need a
suitable temperature.

Experiment 3 To show that a suitable temperature is necessary for germination

1 *Prepare three glass jars as shown in Fig. 8.4. First line the jars with blotting paper and put soil inside.*

2 *Plant three bean seeds, one in each jar. Place the seeds between the glass and the paper, so that you can see what they do. Water them well.*

3 *Place the jars in places where temperature conditions are different:*

- *one outside in the garden in a shaded position, where it is warm.*

- *one on a windowsill inside the classroom where it is very hot.*

- *one in the coolest place you can find. You could put it in a refrigerator.*

4 *Put a thermometer in each jar to measure the temperature.*

5 *Keep watching the seeds for the next two weeks.*

6 *Compare how quickly they germinate.*

7 *Draw a diagram of the experiment.*

8 *Write down what you saw, and what you found out.*

The right conditions for germination

Germination cannot take place unless conditions are right. Seeds need the following conditions before they will germinate. See if you can understand why:

Water

1 Plants usually contain a lot of water. All the chemical reactions taking place inside plants take place in solution in water.

2 If a dry seed does not get water, it cannot start to grow normally.

Air

1 Plants need oxygen from the air for their respiration, especially when it is dark and they cannot photosynthesise.

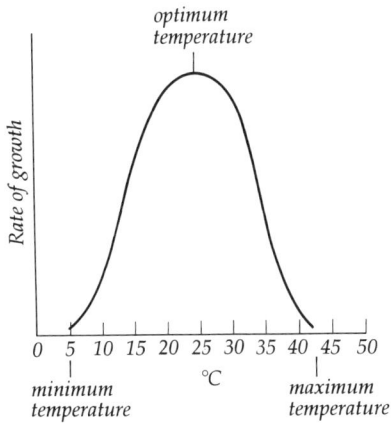

Fig. 8.5 *Temperature in relation to the rate of plant growth.*

2 If they cannot get oxygen, they cannot live very long. Germinating seeds also respire, and they need oxygen.

A suitable temperature

1 Most chemical reactions go faster when it is warm and slower when it is cold.

2 The *rate* of chemical reactions inside a plant is controlled by temperature.

3 Plants cannot control their *own* body temperature, as we can. So their rate of growth depends on the temperature of their surroundings.
This also applies to germinating seeds.

The graph shows how the rate of plant growth changes with temperature. From the graph you can see that:

– The **minimum temperature** for plant growth is about 5 degrees Celsius. No growth takes place below this. As the temperature rises, the rate of growth increases.

– The **optimum** (best) **temperature** for plant growth is between 20 and 25 degrees Celsius. Most plants absorb nutrients, respire and photosynthesise more easily when it is nice and warm.

– The **maximum temperature** for plant growth is about 40 degrees Celsius. This is too hot for most plants. Only cacti and desert plants can survive when it gets as hot as this.

The seed must be alive!

1 This is obvious, but you have to be sure you are not using old seed that has died.

2 Seed that is alive and ready to germinate is said to be 'viable'.

Germination in agriculture

A farmer provides the conditions necessary for germination by ploughing and breaking up the soil to make a fine seedbed.

A good seedbed provides water, air and a suitable temperature so that the seeds will germinate. So remember to make a good seedbed when you want to raise young plants.

Viability of seed

Dry seeds are in a **dormant** state. This means that they are asleep. They are alive, but they are not doing much. Their respiration rate is very slow. The question is 'How long can they stay alive, and how long can you keep them?'

Seeds more than a year old may not germinate. They have 'died in their sleep'! Generally the bigger seeds, such as peas and beans, stay viable (alive) longer than smaller seeds like cabbage, carrot or tomato.

Germination testing

You can check how viable your seed is before you use it, by doing a germination test.

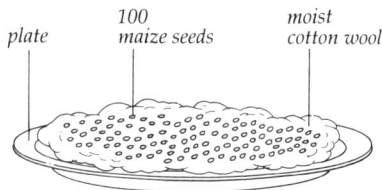

Fig. 8.6 Finding the percentage germination of seeds.

PRACTICAL WORK

Experiment To find the germination percentage of seeds

This means testing to see how many seeds out of a hundred are viable.

1 Put 100 maize seeds on moist cotton wool, exposed to the air. These are good conditions for germination, i.e. the seeds have water, air, and a suitable temperature.

2 Leave them for a week to see how many germinate. Keep a record of their germination in table form, as shown in Fig. 8.7.

Day	1	2	3	4	5	6	7
No. of seeds germinated							

Fig. 8.7 A record of seed germination.

3 Count the seeds that germinate, and the ones that do not. The number out of one hundred that germinate is the **percentage germination** for that sample of seed.

Fig. 8.8 Buy seeds with care.

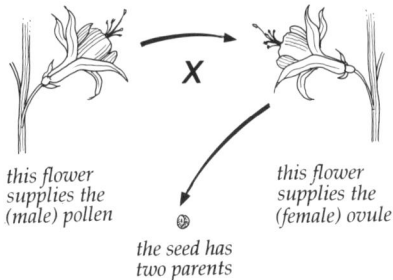

this flower
supplies the
(male) pollen

this flower
supplies the
(female) ovule

the seed has
two parents

Fig. 8.9 A seed has two parents.

4 *When did most of the seeds germinate?*

5 *How do you think that a germination test will assist a farmer?*

6 *What should the farmer do if the germination rate is less than 50%?*

7 *What should you look for when buying seeds?*

Why the sexual process is important

Each seed has two parents. One parent is the plant that produced the pollen, and one is the plant that produced the ovule. The seed was formed when the pollen fertilized the ovule.

A plant grown from a seed has some features of each of its parents. Just as a human child is partly like its father and partly like its mother, so a plant that grows from a seed has some features of each parent.

Notice that the offspring are not exactly the *same* as their parents. **Variation** can occur from one generation to the next. This is important, because it helps plants to survive if conditions get difficult. If they are not all the same, some will always be better fitted to survive than others.

How plants are crossed

Plant breeders can use sexual reproduction to improve crop plants. They choose plants with good features, and purposely **cross** them to combine these features in the next generation.

It is possible to take pollen from the anthers of one parent plant and place it on the stigma of another plant.

This is how to do it:

1 Decide which plant you want to be the female parent, and which one will be the male parent.

2 Cut the anthers out of the female flowers using a pair of fine scissors. Do this as soon as the flowers open, and before the anthers are ripe and able to shed pollen. This prevents the flower from fertilizing itself with its own pollen.

3 Tie a small paper bag on to each flower to keep insects off. They might bring unwanted pollen to

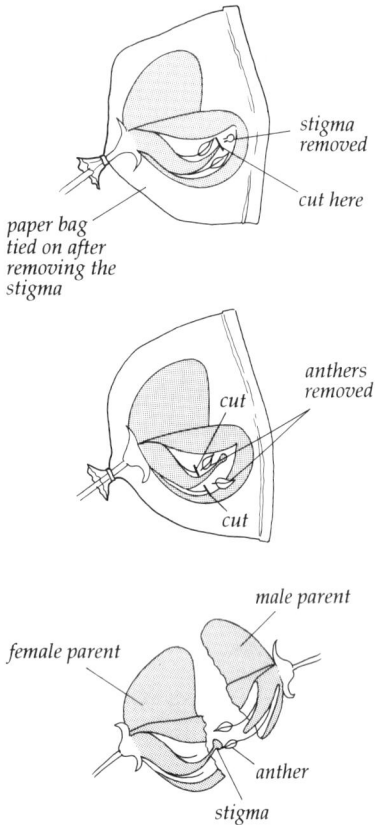

Fig. 8.10 *How to cross-pollinate flowers.*

the flower. You can remove the bags to look at the stigmas, but always put them back.

4 When the stigmas are fully grown and ready to receive pollen, take some pollen from the male parent and put it on to the ripe stigmas of the female parent. A soft paintbrush can be used for this. Cover the flowers again.

5 Cross-fertilization will take place, and seeds will form. You now know exactly which plants are the parents of the seeds.

Crossing plants like this is a good way of improving them. Plant breeders do it all the time. For example, if a potato has large tubers but is not resistant to disease, it can be crossed with another plant that has smaller tubers but *is* resistant to disease.

Some of the potatoes grown from the seeds of this cross may have the good features of both parents – large tubers *and* resistance to disease! *Some* may have the *bad* features of both parents, but these can be thrown away.

Questions

1 (a) Draw a fully labelled diagram of a bean seed cut in half.

   (b) What is the function of the cotyledons?

2 (a) Draw a fully labelled diagram of a maize seed cut in half.

   (b) Where is food stored in this seed?

3 (a) Describe an experiment to show that water is necessary for the germination of seeds.

   (b) Name two other conditions needed for germination, and say why each of them is necessary.

4 (a) Describe an experiment to test the 'viability' of a sample of seed.

   (b) How would you record the number of seeds germinating during one week? Show how you would work out the percentage viability of the seed sample.

5 (a) With the help of diagrams, explain how to cross two related strains of plants.

   (b) Name two advantages that could be gained from crossing the plants.

9

GROWING PLANTS WITHOUT SEEDS

When you have read this chapter you will be able to:

- *explain the meaning of asexual reproduction.*
- *give three examples of asexual reproduction that occur naturally in plants.*
- *give three examples of artificial asexual propagation practised by farmers.*
- *explain why a plant propagated asexually is exactly like its single parent.*
- *describe three advantages of sexual reproduction in plants, and three advantages of asexual reproduction.*

Asexual reproduction

Asexual reproduction means growing new organisms without any sexual process at all. Sometimes it is called **vegetative propagation**. Most animals cannot reproduce in this way, but plants have many ways of doing it. They can grow a whole plant from just a small part of their bodies, and it may be almost *any* part.

Growing whole plants from small parts

Different plants can be **propagated** from different parts. You may already know some of the examples shown in Fig. 9.1 on pages 66 and 67.

Notice how different parts of these plants, (stems, special stems that grow underground, roots and root cuttings), can all be made to grow into complete plants.

Part of the plant	Example	Comments
Aerial stems (part of the shoot) *sugar cane (aerial stem)* *new roots forming at node* *fibrous roots*	Plum trees, sugar cane, and cassava	Pieces of twigs can be planted, and they will take root.
Underground stems: – bulbs *onion (bulb)* *new bulb developing*	Onion Garlic	Onion bulbs grow smaller bulbs on the side. These can be taken off to grow new plants. Garlic bulbs produce many 'cloves'. These can be separated to grow new plants.
– rhizomes *canna lily (rhizome)* *flower stalk* *lateral bud* *stem swollen with food*	Ginger, canna lily, heliconia	A rhizome branches and puts up new shoots. It can be cut up and planted out to get new plants.
– corms *a new corm grows on top of the old one* *dasheen (corm)*	Dasheen, eddoe	The corms can be separated and planted out.
– suckers *a new sucker grows from the base of the plant* *banana (sucker)*	Banana, plantain	Suckers are small plants growing up from the base of the parent plant. They can be cut off and transplanted.

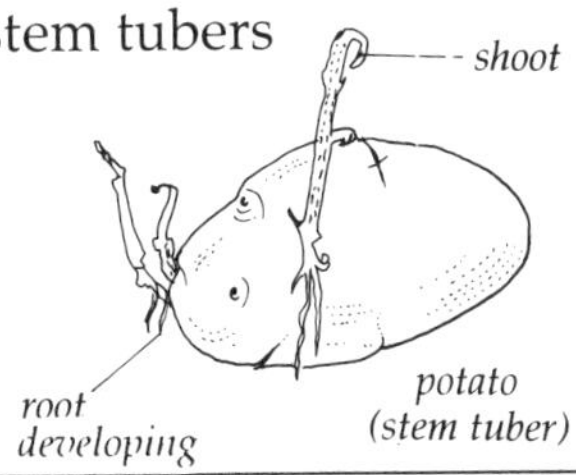 – stem tubers shoot root developing potato (stem tuber)	Yam, potato	Whole tubers of medium size can be planted. Larger tubers may be divided before planting. New plants will grow from these tubers or parts of tubers.
Leaves aloe (leaf)	Aloe	Aloe leaves can be cut off and planted. They grow roots, and produce new plants.
	Begonia	Leaves of large varieties can be pegged to the soil, and will grow new plants.
Root tubers 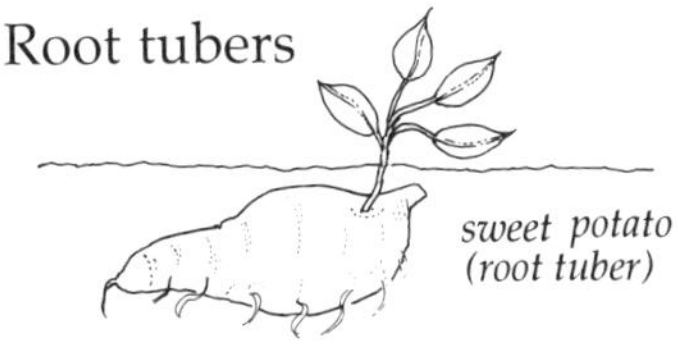 sweet potato (root tuber)	Sweet potato (*Ipomea batatas*)	Whole root tubers of medium size can be planted. Larger root tubers may be divided before planting. New plants will grow from these tubers or parts of tubers.
Root cuttings	Breadfruit, (*Artocarpus altilis*)	A cutting from the root can grow into a new tree.
Bulbils 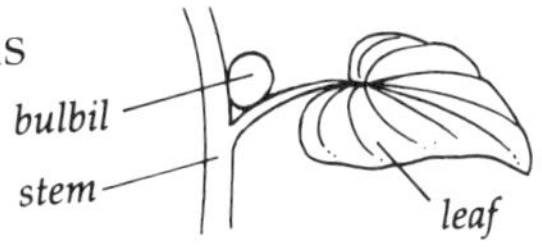 bulbil stem leaf	Yam (*Diascorea alata*)	Axillary buds on the stem of yam plants are modified into **bulbils**. These are fleshy structures stored with food. They fall to the ground and grow into new plants.

Fig. 9.1 Reproduction without seeds.

Notice that some of these examples happen naturally. Others are *made* to happen by the farmer. He can use the ability of plants to grow from small parts by taking a **cutting** of a piece of a plant, and helping it to grow into a whole plant.

For vegetative propagation to be successful, the part of the plant used for propagation must:

- be able to produce a complete root system
- be able to produce a complete shoot system.

To do this it must:

- be alive and healthy
- have enough food stored in it to keep it growing until the new plant can support itself.

Notice that these are the same four conditions that must exist in the seeds. All of them are necessary in the propagation of new plants.

PRACTICAL WORK
Finding examples of asexual reproduction

1 *Go out into the school compound and see if you can find any examples of plants reproducing by asexual means. For example, you might look at potato, banana, cassava, sweet potato, onion and sugar cane.*

2 *Make notes of the ones you find, and bring samples of the smaller ones back to the classroom.*

3 *Draw and describe the methods of asexual reproduction that you have seen.*

4 *Just one example will show you that plants can do something that we and other animals cannot do. They can increase their numbers without breeding.*

A single parent

A plant grown by any asexual method just has *one parent*. In fact it is really part of the very same plant, but it has become separated off. This means that it is exactly like its parent in every way. It has all the same features. Another name for this type of plant is a **clone**. In asexual reproduction, there is no variation from one generation to the next.

This is sometimes useful to a farmer, because a good variety produced sexually, by crossing, can then be propagated asexually from cuttings, tubers, suckers, etc. All the good features of the new variety can be preserved unchanged. Then every new plant will be exactly the same as the first parent. This is a good way of keeping all the features of a good variety of plant.

The story of grapefruit

The grapefruit is an unusual plant. It can produce viable seeds without the normal sexual process. What happens is that the nucellus, inside the ovule of the flower, undergo a kind of vegetative reproduction at a very early stage of development. It divides to produce many nucellar embryos. Each of these can develop into a plant

Fig. 9.2 A grapefruit.

identical to the mother tree. From 60–90% of the plant's seeds may be produced in this way.

This means that if you find a grapefruit that you like and you plant seeds from it, there is a very good chance that you will get trees just like the single parent. The only trouble is that the *rest* of the seeds are produced by the normal sexual method involving cross-pollination and variation, so you cannot be sure that *all* your trees will be the same.

Commercial citrus growers do not rely on seeds. They prefer to propagate their trees by **budding**. This method of propagation will be discussed in Year 3.

Comparing sexual and asexual reproduction

The question is, 'Which is the best way for plants to reproduce?' – and the answer is, 'There are advantages in both methods.'

Look at the table below. See if you can see why both methods of reproduction can be useful to both plants and farmers.

ASEXUAL REPRODUCTION	SEXUAL REPRODUCTION
Advantages	*Disadvantages*
Young plants can get food from the parent plant as they grow.	The seed has only a small amount of food stored in it.
They find themselves in good conditions, where parent plants are already growing.	Some seeds fall in places where they cannot germinate. They need soil, water and a suitable temperature.
Good features can be preserved by this method.	Variation due to sexual reproduction is not always what the farmer wants.
Disadvantages	*Advantages*
Plants may be overcrowded when they all grow close to their parent.	Seeds can spread over a wide area to spread the plants about.

No new varieties can be produced.	New varieties can arise by cross-breeding. Plant breeders can use this method.
It is not easy to grow large numbers by asexual methods.	Large numbers can easily be grown from seeds.
Since all plants have the same genetic make up, pests and diseases may spread quickly from plant to plant.	Genetic differences mean resistance to the spread of pests and diseases is greater.

Questions

1 (a) What is 'asexual reproduction'?

 (b) Name three crop plants that can be propagated vegetatively, (asexually). In each case describe how the plant material is prepared for propagation and how it is planted. You may use diagrams in your answer.

2 (a) What must the part of a plant used for propagation be able to do?

 (b) What two qualities are required in this part of the plant to make it suitable for propagation?

3 How can a plant breeder:

 (a) produce a new variety of potato?

 (b) propagate the new variety so that all the tubers are identical?

4 (a) A plant propagated by vegetative means has just one parent. Give two reasons why this can be an advantage to the plant.

 (b) Describe two possible disadvantages.

5 A grapefruit tree can produce some seeds by asexual means and some by sexual means.

 (a) What kind of tree would you expect to grow from a seed produced asexually? Give reasons for your answer.

 (b) What kind of tree would you expect to grow from a seed produced sexually? Give reasons for your answer.

10

TOOLS AND EQUIPMENT

When you have read this chapter you will be able to:

* *recognize and name various garden tools.*
* *explain the uses of each of them.*
* *use the tools with confidence and skill.*
* *keep tools clean, and look after them carefully.*
* *use all tools safely.*

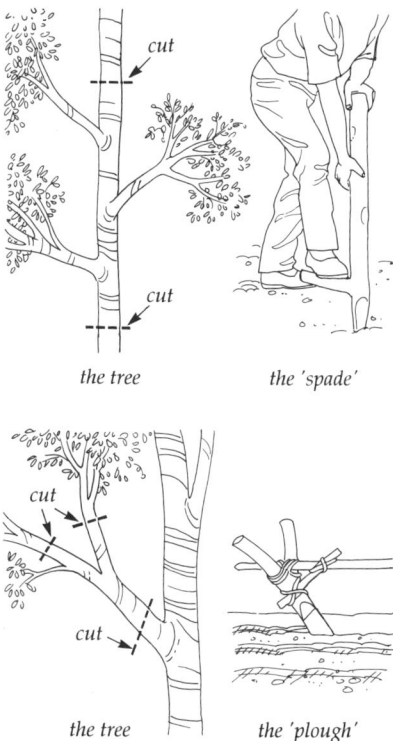

the tree   *the 'spade'*

the tree   *the 'plough'*

Fig. 10.1 *Tools made from pieces of trees.*

How tools are made

The first people to farm just had 'digging sticks' to break up the soil and plant their seeds. Some people still use wooden implements.

You could try making a wooden 'spade', just once, to see if it works. The secret would be to find the right tree, where you could see the shape of the tool in the branches.

Can you see how pieces of trees can be made into tools? Do you think they would work very well? They probably did not work very well for our ancestors either! If you tried to make a wooden spade, you would soon realize what good tools we use today!

Modern tools make work easier. The tools we use now are:

– well designed, for different jobs
– strong
– light in weight
– hard wearing
– suited to the height and strength of a person.

trowel

watering can

hand fork

file

bush knife

sprayer

screwdriver

hoe

fork

spade

slasher

rake

hammer

saw

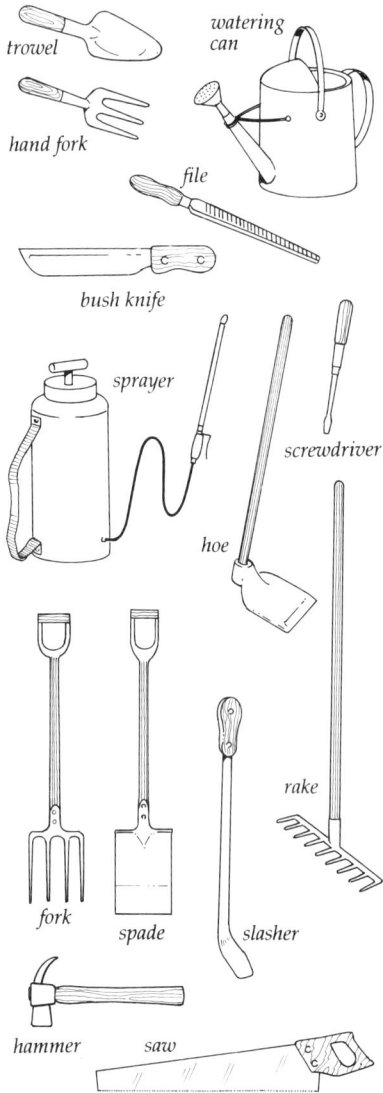

Fig. 10.2 *Garden tools. Can you say what each tool is used for?*

Fig. 10.3 *Pendulums.*

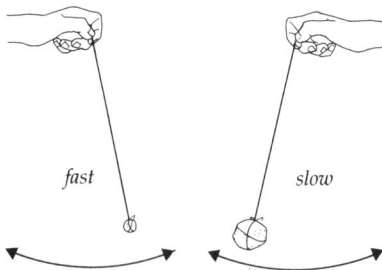

fast

slow

a) The small stone swings more quickly than the big one.

Exercise

How different tools are used

1 Do you know the names of all the tools?

2 Can you say what each one is used for?

3 Make sentences which make sense from the half-sentences in this table:

The soil is dug	with a rake
Weeds are cut down	with a watering can
Soil is levelled	with a spade
A straight edge is made	with a brushing cutlass
Plants are watered	with a line

Gaining skill with tools

The weight and length of a tool influences its movement. When we swing a hammer or a hoe, we can feel this.

If you hang a small, *light* stone from a length of string and let it swing, it moves back and forwards quite *quickly*. If you hang a *heavy* stone from a string of the same length, it moves *slowly*. If you shorten the string, the movement becomes faster. Try it and see. The speed and rhythm of a moving **pendulum** is constant for its weight and length. That is why grandfather clocks all had a pendulum. The speed of their swing never changed, so it could be used to measure time.

A light tool like a hammer will swing more quickly than a heavy hoe. Also, the shorter the tool, the faster it swings; and the longer it is, the slower it goes.

With practice you will grow more skilled in using tools. The secret is to let their weight and movement do the work for you. You can try this out in the garden. First use a hoe at normal speed, then go slow, then go fast. You will soon find the 'normal' speed for the hoe. If you don't do this, you will waste your energy. If you cooperate with the hoe, you can do better work, and carry on for longer.

pendulum

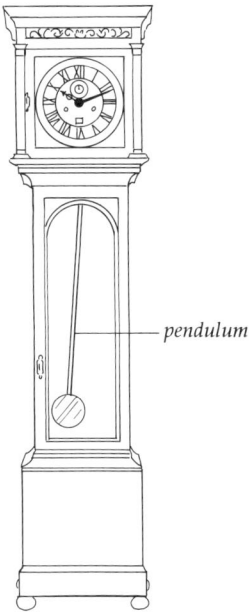

b) The pendulum inside a grandfather clock.

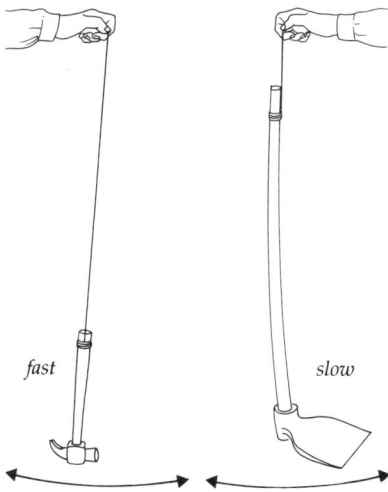

fast

slow

Fig 10.4 *Swinging a hammer and swinging a hoe.*

Practice makes the work easier, because your skill will improve. You will soon be able to use the tools safely and confidently.

Some field workers have 'hoeing songs'. They sing as they work! The rhythm of the song fits the rhythm of the hoe, and singing helps them to keep the work going on steadily.

Care of tools

1 Always clean tools after you use them.

2 If you are not going to use a tool for some time, put some oil on a piece of cloth and wipe the blade with it. This will stop it from going rusty (Fig. 10.7).

3 Look after tools properly, and they will last for years.

Using tools safely

It is very important to learn to use tools *safely*. You can easily hurt yourself or others, even when you do not mean to (Fig. 10.8). Tools have no mercy! These are the rules for using them safely:

Using tools safely

- Carry tools with the blade down. Never carry them over your shoulder. Why?

- Do not turn round suddenly when carrying a tool. Why?

- Make sure that you have room to work. Do not let anyone stand too close. Why not?

- Keep your hands out of the way! If you are holding a hammer in one hand and a nail in the other, be very careful not to hurt yourself.

- Keep your feet out of the way when you use a fork or a spade. Why?

Things to do

1 *Compose a poem, song or calypso to tell others about using tools safely.*

2 *Draw a poster or comic strip to illustrate safe use of tools.*

Fig. 10.5 Which tool moves faster?

Fig. 10.6 A hoeing song.

Fig. 10.7 How to prevent rust.

Farm machines

Machines (Fig. 10.9) can make farm work a lot easier. They can speed up jobs such as planting, and weeding between the rows of plants. Some are designed to be pushed by hand, but the bigger ones are powered by a petrol engine.

Tractor-mounted implements

Tractors can do work a lot quicker than people or animals. They can be fitted with attachments for most field operations, e.g. ploughing, harrowing, disc harrowing, planting, ridging, cutting grass, spraying crops, etc. Tractors can move about quickly and easily, and lower or raise their implements as required, using their hydraulic lift.

Tractors can also supply power, through a pulley, or through a power take-off shaft to drive machines that stand still. Machines such as maize mills, or irrigation pumps, can be driven in this way (Fig. 10.11).

The cost of tools and machinery

With good tools and powerful machines, farming is made much easier. But such things cost a lot of money! A farmer has to be very sure he is going to get his money back from increased production before he buys a new machine.

The questions he asks himself are:

- 'How much will this machine cost to buy?'
- 'How much will I have to spend on servicing, repairs and fuel?'
- 'Will I have to pay less for labour?'
- 'How much less?'
- 'Is it worth it?'

Until he knows the answers to these questions he had better not buy any machine. The big question is, 'Will it pay?'

SAFETY FIRST!

Be careful how you carry your rake . . .

. . . Someone may be walking behind you!

Fig. 10.8 *Using tools safely.*

Fig. 10.9 *Some small farm machines.*

a) A rotary cultivator used to prepare seedbeds in one operation.

b) A two-wheeled cultivator.

Fig. 10.10 *A tractor-drawn implement.*

Questions

1 (a) List five rules for using garden tools safely.

   (b) After using tools in the garden, how would you clean and store them?

2 (a) Draw three different garden tools used for digging.

   (b) Explain how they are designed for the work they have to do.

3 What must a farmer consider before deciding to mechanize the work on his farm? Explain how he could gain, and how he could lose, by mechanizing.

4 (a) Name three tractor-mounted implements.

   (b) Explain how each of these implements does its work quicker and easier than the hand tool it replaces.

5 (a) What does the 'hydraulic lift' on a tractor do?

   (b) Why is it better at controlling implements than a simple draw bar?

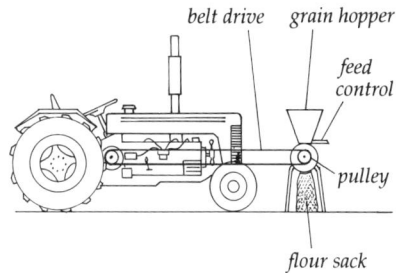

belt drive   grain hopper   feed control   pulley   flour sack

Fig. 10.11 *Using a tractor to drive a maize mill.*

11

CULTIVATION OF A FIELD CROP

When you have read this chapter you will be able to:

- *choose a crop to grow from seed.*
- *prepare a plot for planting, by:*
    - *- digging and manuring the soil*
    - *- producing a fine tilth for seed sowing.*
- *plant the seeds at the right spacing and depth.*
- *look after the crop until maturity.*
- *harvest the crop when it is ready.*
- *use your produce to cook something nice to eat.*

PRACTICAL WORK

How to grow a direct-seeded crop

As part of your agricultural course, you may be expected to plant one of the crops listed below:

- Corn   – *sweet corn, field corn, or a hybrid variety.*
- String beans – *non-vining or vining beans.*
- Bodi bean   – *a dwarf or a vining variety.*
- Ochro (or okra) – *local varieties or imported hybrids.*

All these crops are grown from seeds. They are called **direct-seeded crops** *because there is no need for*

Fig. 11.1 Comparing weeding with not weeding.

transplanting, and no vegetative propagation involved. You just sow the seeds directly into the plot and let the crop grow to maturity. Your teacher will help you.

To make the most of this exercise you can try different treatments on the crop. Each group in the class could try a different treatment on a part of their plot. Some suggestions are given below. Try out and compare the results of these alternatives.

• manuring the land	• not manuring it
• fertilizing	• not fertilizing
• planting early	• planting late
• close spacings of rows	• wide spacing of rows
• weeding	• not weeding
• top dressing	• no top dressing
• irrigation during drought	• no irrigation

As the crop grows, you will see it respond to the conditions provided for it. You will see which conditions it prefers. Watch carefully, and record anything that seems interesting. At harvest time, compare the yields from the different plots to see which treatments worked best.

Costing

A farmer always thinks about costs. He wants to know whether a particular treatment will increase the yield of the crop, and whether it will pay for itself. So, when you grow your crop, ask:

> 'How much did this treatment cost for a plot the size of ours?', and 'How much would this treatment cost per hectare?'

At harvest time, ask:

> 'Did the treatment increase the yield?'
> 'Did it increase the profit?'
> 'Was it worthwhile?'

Drawing conclusions

At the end of the growing season, compare with your classmates all the treatments tried out by the class. Try

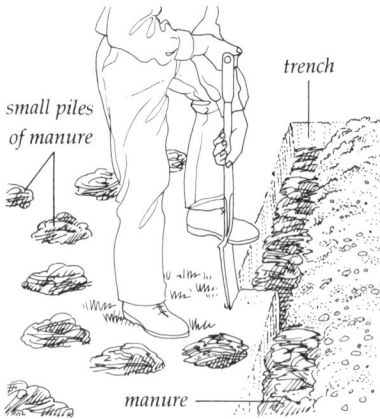

Fig. 11.2 *Incorporating manure.*

Fig. 11.3 *Preparing a seedbed.*

to work out the most successful combination of practices for growing the crop. Write a report on the whole exercise, based on notes and observations you made when the crop was growing. Say what you did, and what you think were the best treatments for the crop.

Land preparation

The following suggestions will guide you. Keep checking to see that everything has been done.

1   Clear the plot of land of all weeds and put the weeds on the compost heap.

2   Remove any large stones and rubble from the plot.

3   Till (dig) the soil using spades or hoes, and if you are applying manure or compost, dig it in. Fertilizer and agricultural lime could also be added at this stage. Your teacher can tell you how much to put on.

4   After primary tillage, break down any large clods with a fork or rake to produce a fine tilth. Remember that seeds need good conditions in which to germinate. They need:

   – **Water** to enable the dry, dormant seeds to swell up and start growing.

   – **Oxygen** in the air, (which is in the spaces between the soil particles), to let the seeds 'breathe'.

   – A **suitable temperature** – not too hot, and not too cold.

A seedbed with a fine tilth provides these conditions. When the rain falls the seeds will begin to grow.

5   In areas where heavy rains could be a problem, dig drains 30 cm to 40 cm wide and form beds 90 cm to 120 cm wide × 1.8 m to 2 m long for each group.

Planting the crop

1   Use disease-free seeds from a previous crop or seeds purchased from a reputable supplier.

2   Planting distances for the different crops are as follows. Use these measurements unless you are deliberately experimenting with alternative spacings.

surface of soil

sowing depth is
five times
diameter of seed

seed

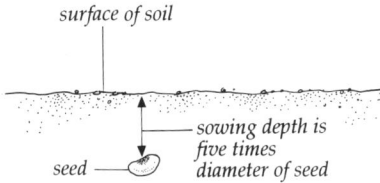

Fig. 11.4 *How to find the right depth for sowing seeds.*

Crop	Within row	Between row
Corn	10 cm to 30 cm	60 cm to 100 cm
Ochro	60 cm to 75 m	60 cm to 75 cm
Bodi	60 cm to 75 cm	60 cm to 75 cm
String bean	15 cm to 30 m	15 cm to 30 cm

Set lines at the right distance apart, using a measuring rod. Then draw drills, following the lines, at the right depth for sowing the seeds. Sow the seeds at the recommended spacing along the drills. Cover them with soil.

3 *Check what is the best time of year for planting your crop in your particular area. Stick to the right time, unless you are experimenting with planting times.*

Looking after the crop

*The main things to be done are **weeding, controlling** any **pests** and **diseases**, and carrying out any **cultural practices** such as pruning, staking, moulding, mulching. Carry out the necessary operations, and keep a record in your diary of what you did, and the dates.*

Weeding

A weed is a plant growing in the wrong place. Weeds will compete with your crop and take as much as they can of:

- *plant nutrients from the soil*
- *water from the soil*
- *space in the plot*
- *light.*

You work hard to prepare the land and provide good conditions for your crop, not for the weeds! They will try to take advantage of your hard work. Also, they may carry pests and diseases that can spread to the crop, and they make the garden look untidy.

Weeds are the enemy! Cut them down with hoes between the rows. Pull them up by hand where they grow between the plants in the rows. Put them on the compost heap where they will decompose (rot) and make useful humus.

Fig. 11.5 *Getting rid of weeds.*

Fig. 11.6 *Checking for pests and diseases.*

One important point – never let weeds live long enough to flower and set seeds. *If their seeds fall on the ground, there will be even more of them next year. One of the best ways of controlling weeds is to prevent them from seeding.*

Pest and disease control

Look at the leaves of your plants from time to time. Turn them over to see if insects are living on them or eating them. Also look for any coloured spots on the leaves, caused by fungal disease. Check the flowers and pods. If you see any signs of pests or diseases, tell your teacher straightaway.

The larger pests can be picked off by hand and crushed under your foot. Spraying may be necessary to control some pests and diseases. Your teacher will show you how to do this. Students should never try to use chemical sprays. Let your teacher or the farm attendant spray your crops. Sprays can be dangerous.

Cultural practices

If your crop requires pruning, staking or mulching, check with your teacher when to carry out these operations.

Maturity and harvesting

The approximate times from planting to harvest for the different crops are:

Corn	3–5 months
Bodi	6–11 weeks
Ochro	8–10 weeks
String beans	6–7 weeks

Experiment with harvesting at intervals to find the best time for harvesting. You will soon recognize when the crop is ready. Weigh all produce and record the yield.

Try preparing something nice to eat from your crop. This will be your reward for growing it! Find a recipe and have a good meal.

Storing and preserving the crop

Can your produce be stored or preserved to make it last a long time? Find out the best way of preserving it.

Questions

1 (a) What is a 'direct-seeded crop'?

   (b) Name one such crop, and write brief notes on how it is cultivated. Include:
- land preparation
- fertilizing
- sowing
- spacing of rows
- after cultivations
- harvesting.

2 (a) Describe the inputs needed for a named field crop.

   (b) How can you work out the cost of these inputs?

   (c) After the harvest, how can you work out the total returns from the crop?

3 (a) What is meant by 'weed competition'?

   (b) How can you help your crops to win the battle against weeds?

4 (a) Name four signs that tell you that pest or disease organisms are attacking your crop.

   (b) Describe three 'cultural control' measures for dealing with pests and diseases.

   (c) Give two reasons why cultural control is better than using chemical sprays.

5 (a) Name two dangers of 'agro-chemicals'.

   (b) Describe the safety precautions you should take when using chemical sprays.

12

LIVESTOCK MANAGEMENT

When you have read this chapter you will be able to:

- *say what is meant by livestock management.*
- *list the main groups of vertebrate animals.*
- *describe the classes of livestock kept on farms.*
- *explain why each is kept, and what it produces.*
- *say what is meant by selective breeding.*
- *show why domesticated livestock are more productive than their wild relatives.*

The place of farm livestock in the animal kingdom

Scientists classify animals, (put them into groups), by how closely they are related to each other. This is a natural way to group them. Animals in the same group look similar to each other because they come from common ancestors. The whole **animal kingdom** is classified in this way.

There are two main divisions in the animal kingdom – the **invertebrates** and the **vertebrates**.

Invertebrates

The most ancient groups of animals have no backbone. They are the invertebrates, and include animals like worms, snails, insects, spiders, crabs and centipedes. They are not usually kept as farm animals, though in some parts of the world there are shrimp farms, and in

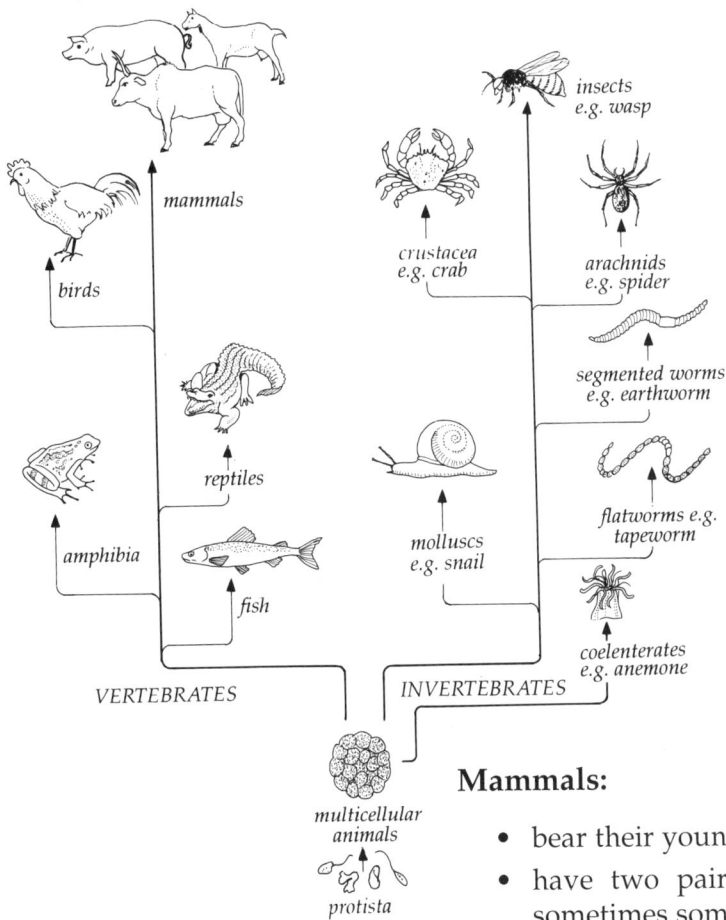

Fig. 12.1 The animal kingdom. Can you find the groups to which some common farm animals belong?

France certain kinds of snails are kept for eating. We make use of bees, which are insects, to help pollinate fruit trees.

Vertebrates

Animals that have a backbone are called vertebrates. Within this group there are five **classes**. These are **fish**, **amphibians**, **reptiles**, **birds** and **mammals**. Do you know to which group we ourselves belong?

Most farm livestock are either mammals or birds. Members of these groups have certain features in common.

Mammals:

- bear their young alive
- have two pairs of five-fingered limbs, though sometimes some of the 'fingers' are missing
- have bodies covered with hair
- suckle their young with milk
- have a warm body temperature.

Birds:

- lay eggs, and **incubate** them by keeping them warm until the young hatch out
- walk on two hind limbs, which are covered in scales, (birds are thought to have descended from reptiles)
- have their fore limbs modified to form wings
- have bodies covered by feathers, which are modified scales
- have a warm body temperature.

Because we are mammals, we know what it feels like to be a mammal. This should help us to understand how

Fig. 12.2 *External features of farm animals.*

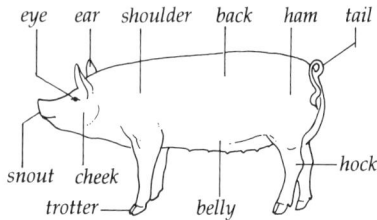

a) The pig

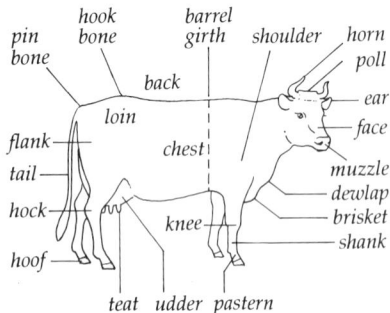

b) The cow

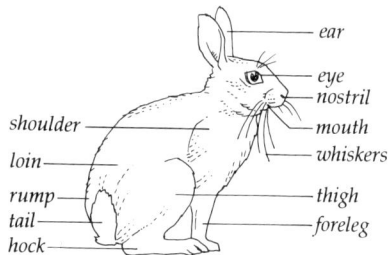

c) The rabbit

mammals should be looked after. Their needs are similar to ours.

Most domesticated birds do not fly very well and prefer to stay where they can get food, rather than try to escape. We keep them for their eggs and their meat.

The uses of farm animals

From the time that people first took animals from the wild and domesticated them, they took responsibility for managing them and meeting their basic needs. This was a small price to pay for the advantage of having a source of protein food readily available.

What is meant by 'livestock management'?

One way to answer this question is to discuss it in class, and decide what it means. Try this now.

1 Divide the class into groups. Let each group select a leader, a secretary (note-taker), and a reporter.

2 Discuss the topic Livestock Management for 15 minutes.

3 Make a list of as many farm animals as you can think of. Now classify your list of animals into groups, (poultry, large farm livestock, smaller farm livestock). Discuss how each group is managed and looked after. For each group, you could discuss:

   – types, breeds or strains of animal
   – purpose for rearing the animals, i.e. what you get from keeping them
   – housing requirements
   – equipment requirements
   – reproduction/breeding of the animals
   – care of the young
   – sanitation
   – food and water requirements
   – prevention and control of diseases and pests (bacteria, viruses, deficiency diseases, parasites, vaccines, antibiotics, disease resistance)
   – protection from predators
   – keeping records.

4 Do not worry if you do not know everything yet. You will be learning more in this chapter – but discussing the subject helps you to realize what is involved in keeping animals. Make notes of the ideas and opinions of the group.

5 The group 'reporter' can give a short talk to the class on what the group said about Livestock Management. The written report from the group can be given to your teacher at the end.

6 Your teacher can summarize the results of the discussion, and the reports of the groups, on the chalkboard, including:

   • the main points that a farmer must take into account when keeping farm animals.

   • any pattern in the requirements of the different classes of livestock, e.g.
     – in what ways their needs are the same.
     – in what ways they are different.

Collecting information about livestock

1 Make a labelled chart or a scrapbook using photographs or drawings of farm animals taken from magazines, or drawn by you.

2 Make a table summarizing the methods of caring for each class of livestock in your chart or scrapbook.

Classifying farm animals

Scientists may classify animals according to their ancestry, but farmers usually group their livestock according to their size, and what they produce.

MAMMALS

Large livestock	Kept for
beef cattle	meat
dairy cattle	milk, mozzarella cheese for pizza
horses	riding, pulling farm implements
donkeys	riding, pulling farm implements
water buffalo	pulling farm implements
buffalypso	meat

Fig. 12.3 *How farmers group their animals.*

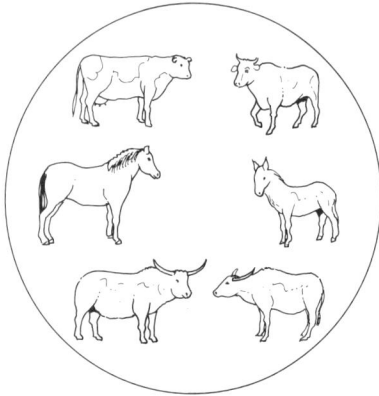

large livestock

smaller livestock

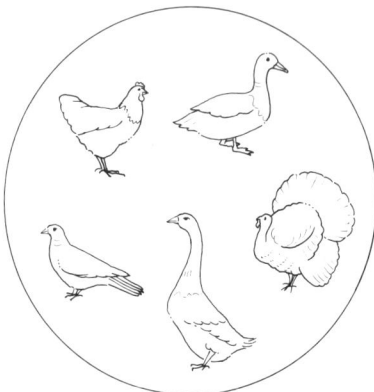

poultry

The buffalypso was produced by selective breeding of the water buffalo. This process started in 1949, and now the buffalypso animals are much more solidly built than the buffalo, and produce very good quality meat. In Trinidad and Tobago alone there are about 6000 buffalypso.

Smaller livestock	Kept for
sheep	meat, sometimes milk
goats	meat, milk
pigs	meat
rabbits	meat, pelts

BIRDS	
Class of poultry	Kept for
chickens	eggs, meat
turkeys	meat
ducks	eggs, meat
geese	eggs, meat, protection
pigeons	meat, sport racing

The needs of the larger animals are different from those of the smaller livestock:

1 Large animals need more space, bigger housing, stronger fencing and more food. They live longer than small stock, and take longer to reach maturity and start producing something. This means that the farmer who keeps large animals has to wait a long time for a return on money invested.

2 Smaller stock need less space, are easier to house, have a shorter life cycle and give a quicker return on money invested.

Later, we shall study methods of looking after the different classes of livestock in more detail. For the moment, try to remember their names, and the reasons why they are kept.

Other uses of animals

People keep animals for food or valuable raw materials and other reasons as well. In the table on p. 87 some uses of animals are listed. See how many examples of

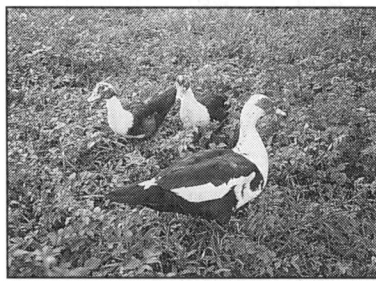

Fig. 12.4 *To which groups do these animals belong?*

animals you can think of that are kept for these reasons. There are a lot of them!

Copy and complete the table:

Animals used for	Examples
1 Power for lifting, carrying or pulling	
2 Guarding people and property	
3 Keeping as pets	
4 Hunting	
5 Riding	
6 Food	

Improvement by breeding

Animals have changed by being domesticated. Ever since farmers first began to keep livestock, they have chosen their 'best' animals to be the parents of the next generation. They would tend to choose the tamest, or the biggest, or the best milkers, depending on what they wanted their animals to be like. This is what you would expect them to do.

This kind of 'selective breeding' has been going on for many generations. It has greatly influenced the way in which farm animals have developed, and it explains why they are now so much more productive than their wild relatives.

Today, scientists understand better than ever before how to breed animals systematically, and our domestic livestock are still being improved. Animals from different breeds and even from different parts of the world are now crossed to combine their good qualities.

Can you think of any examples of this being done in the Caribbean?

The Pig
The pig
Is very
Very
Big.
And sometimes even bigger;
For all that swallowing
And wallowing
And excessive hollering
Do nothing at all
However small
To improve his figure.

Frank Collymore

Questions

1 (a) Name two classes of vertebrates to which farm animals belong.

   (b) To which class of vertebrates do human beings belong?

   (c) Describe four features of a cow that show it belongs to the same class as ourselves.

2 (a) List four types of poultry.

   (b) What useful products are obtained from each type?

   (c) Why are domesticated birds more productive than wild ones?

3 (a) What is meant by 'selective breeding'?

   (b) How could selective breeding improve milk production?

4 (a) Name two types of invertebrate animals kept by man.

   (b) For each type you have named, explain briefly how the animals are kept, and what products are obtained from them.

5 (a) What is a 'buffalypso'?

   (b) How was it produced?

   (c) Why is this a useful animal?

13

KEEPING POULTRY

When you have read this chapter you will be able to:
- *list five different types of poultry.*
- *draw a chicken and label it to show the parts.*
- *list five breeds of chickens.*
- *name the broiler and the layer breeds.*
- *explain the uses of poultry.*
- *list the characteristics of a good broiler.*
- *explain how to build a house for chickens.*
- *describe three different poultry production systems:*
     - an extensive system
     - a semi-intensive system
     - an intensive system.
- *rear a batch of broiler chickens, including:*
     - preparing the poultry house
     - equipment needed
     - preparing for the arrival of day-old chicks
     - brooding
     - vaccination and debeaking
     - routines for looking after broilers
     - feeding and watering
     - keeping records
     - cleaning out the poultry house between batches
     - disease prevention and control
     - marketing of broilers
     - how to 'dress' a chicken.
- *carry out a survey of nearby poultry enterprises.*
- *explain what is meant by breeding and incubation.*
- *demonstrate how to make a brooder.*
- *demonstrate how to build a poultry house.*

Fig. 13.1 *Some types of poultry.*

a) *Chickens*

b) *Turkeys*

c) *Geese*

d) *Guinea fowl*

Poultry are birds that have been domesticated by people. Several types of birds are included in this class of livestock. Do you recognize them from the pictures?

Collecting information about poultry

1 Make a list of as many types of poultry as you can think of.

2 Make a labelled chart or a scrapbook using photographs or line drawings of poultry animals. You could get photocopies from books in your school or public libraries, and collect pictures from magazines.

3 List the reasons why people rear poultry.

4 If you can, collect pictures of different breeds, or strains, of each type of poultry.

Rearing a batch of poultry

The best way to learn about poultry is to rear a batch of birds yourself. Before you start you will have to make several decisions. Here are some points for you to discuss in class and make decisions about:

– What type of poultry would you keep?
– What type of housing should you provide?
– What breed of birds should you choose?
– What should you feed them on?
– Where will you obtain your stock?
– How much will it cost to keep them?

This chapter concentrates mainly on chickens. They are the most common type of poultry kept on farms, and are probably best for keeping in schools.

Parts of a chicken

These are shown by Fig. 13.2.

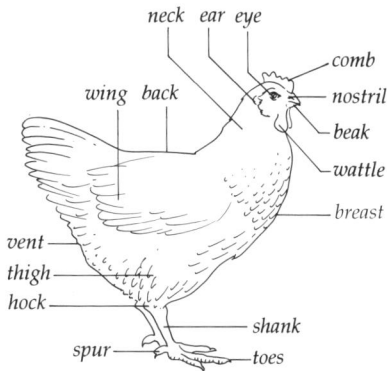

Fig. 13.2 The parts of a chicken.

Fig. 13.3 Some common breeds of chicken.

a) Plymouth Rock

b) White Leghorn

Breeds of chickens

Many breeds of chickens have been developed in different places. Here are some of them.

Pure broiler breeds	Broiler hybrids	Layer breeds
Plymouth Rock	Vantress Cross	White Leghorn
Rhode Island Red	Golden Cobb	Rhode Island Red (dual purpose)
	Videttes	

Each breed has its own special qualities. Note that it is also common practice to cross different breeds to produce **hybrid** birds for particular purposes. **Broilers**, for example, are often hybrids. They have been specially bred to put on weight quickly to provide chicken meat for the market.

Uses of chickens

Chickens are kept for producing eggs and meat, but different breeds and crosses are used for each purpose, as follows:

- **Layers** are kept for producing eggs.
- **Dual purpose** birds are kept for producing eggs, and after their laying life they are eaten.
- **Broilers** are kept for meat production. They grow very fast and are ready for slaughter at the age of about eight to ten weeks. The **broiler hybrids**, (birds specially bred for quick growth), give a quicker return on money invested than the dual purpose 'pure' breeds.
- **Breeding flocks** are kept to produce eggs for hatching. They supply the poultry industry with day-old chicks.

Where to begin?

If you are just starting to learn about poultry production, begin by keeping a batch of broilers. You can get results within a school term, and you will learn a lot from the experience. (See page 98 for details of how to look after broilers.)

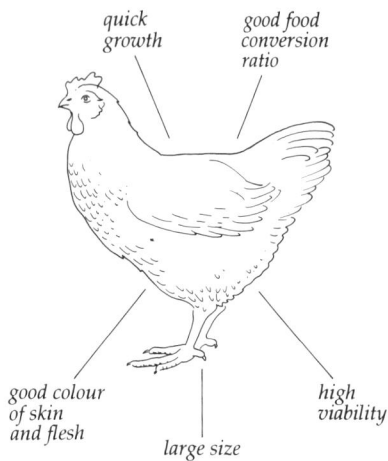

Fig. 13.4 A good broiler chicken.

Reasons for keeping broilers

Farmers keep broilers for several reasons:

1 *Broilers are easy to sell.* People like to eat chicken meat. It is a good source of protein in the diet, and cheaper than other kinds of meat.

2 *Only a small area of land is needed* for broilers. A small broiler house can produce a lot of birds for sale.

3 *It does not cost much* to build a broiler house and buy the equipment needed for keeping the birds.

4 *It does not take long to get some money back.* The broilers are sold at about ten weeks old.

5 *Several batches* of broilers can be kept in one year.

What makes a good broiler?

A farmer looks for these things in a good broiler:

1 **Quick growth**. The bird must put on a lot of weight in the first eight weeks of its life. If it grows fast, it takes less time, less labour, and less food to produce a meat carcass.

2 **A good food conversion ratio** of about 2:1. This means that if the broiler eats 2 kg of food, it puts on 1 kg of body weight.

3 A **good skin** and **flesh colour**. Some customers like a yellow skin. Others like a white skin. This depends partly on the breed, but if broilers are given feed containing yellow maize, the skin will have a yellow colour.

4 A **good size**, and **plenty of flesh**. Then customers will want to buy it.

5 A **high viability**. This means that not many of the birds die when they are small. If birds die, there is less profit for the farmer.

Housing for chickens

The houses where you keep your birds should give them:

- protection from bad weather
- protection from predators.

---

a)

b)

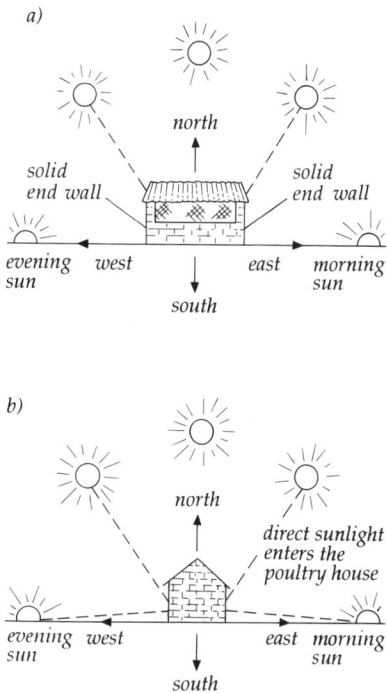

Fig. 13.5 *Which diagram shows the best position for a poultry house?*

This is especially important for young chicks which are not yet fully feathered, as these can suffer from variations in temperature during the day. They can become too cold, and they can be burned by the direct rays of the sun. At night they may try to keep themselves warm by huddling together, and then some may die from crushing or suffocation.

Older birds can also suffer from heat stress, which causes them to drink large quantities of water and not eat enough food. If they are broilers they do not grow fast enough, and if they are layers they may stop laying.

Buildings in the West Indies should be designed to cope with these problems. The first thing to do when you build a chicken house is to make sure that it is aligned east-to-west and not north-to-south. This alignment reduces the amount of direct sunlight entering the chicken house, and so reduces the tendency for its temperature to rise. It also reduces the chilling effect of the Trade Winds. The roof should be made with overhanging eaves to protect from driving rain as well as from direct sunlight. Some farmers hang empty feed bags on the wire netting on the windward side of the poultry house when it is necessary to protect the birds.

Poultry production systems

You may ask why chickens need any special housing and protection at all? *Wild* birds manage alright, so why should chickens not be the same? The answer is that they are not good fliers, and they are a bit slow. Dogs or thieves would soon catch up with them. This is why they need protection from **predators**.

There are three main systems of keeping poultry:

- the extensive system
- semi-intensive systems
- intensive systems.

The extensive system

Under this system, chickens are allowed to roam freely, and find most of their own food. They lay their eggs where they choose. This is sometimes called the **free range** system. If they survive to become adult birds, they can look after themselves quite well. This system has good points and bad points:

Good points	Bad points
1 You do not have to provide any housing or fencing, so this is a cheap system. The chickens can roost in trees or bushes.	1 Chickens may get into the garden and spoil vegetables or crops.
	2 There is little protection from predators or thieves.
2 There is not much work involved, except for finding and collecting the eggs.	3 You cannot be sure that the birds are getting a balanced diet.
	4 The birds do not produce as well as they might.

Semi-intensive systems

In these systems the idea is to provide just enough housing and protection for the chickens, but not to spend too much money. Poultry coops are placed in a fenced area, so that the chickens can run about and find some of their own food, but they can be shut in at night to keep them safe.

Another method is to keep the chickens in moveable **folds** or **arks** that can be shifted to new ground every few days.

Again there are good and bad features of this method:

Good points	Bad points
1 The chickens are protected and controlled.	1 The fencing and the folds have to be provided, and kept in good repair.
2 They can be given a well balanced diet.	

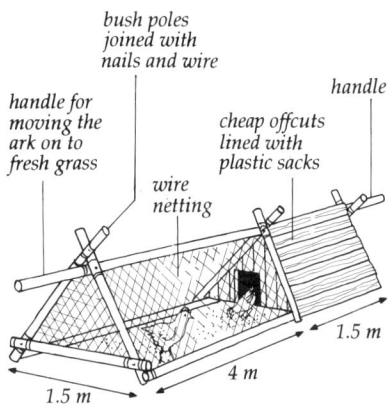

bush poles joined with nails and wire

handle for moving the ark on to fresh grass

cheap offcuts lined with plastic sacks

handle

wire netting

1.5 m

4 m

1.5 m

Fig. 13.6 A semi-intensive poultry ark or fold (moveable).

Fig. 13.7 *Broiler chickens kept on deep litter.*

Fig. 13.8 *Laying hens being kept on deep litter. Note the nest boxes for egg laying.*

3 You can check that they are healthy and free from disease.

4 The level of production is higher than under the extensive system.

2 There is more work involved in looking after the chickens every day.

3 This system costs more to run.

Intensive systems

In these systems, the birds are kept inside all the time. There are two main types:

- the **deep litter** system
- the **battery** system.

The deep litter system

In this system, the birds are kept inside a chicken house, and the floor is covered by some kind of material that will absorb their droppings. Both laying hens and broiler chickens can be kept in this way. Laying hens need to be provided with nest boxes where they can lay their eggs.

A thin 'litter' of dry grass, chopped straw, wood shavings or bagasse from the sugar cane factory, is first put on the floor before the birds arrive. If very young chicks are to be kept, this first layer must not be too deep, or the chicks might get lost in it. More litter is added to keep the floor looking clean, so that the depth of the litter gradually builds up.

The birds seem to enjoy scratching about in the litter, and as they do this their droppings mix with the litter and break down by bacterial action. Heat is given off which keeps the litter dry. Deep litter houses should be well ventilated to get rid of ammonia gas and water vapour given off by the litter.

The battery system

In this system, laying hens are kept in rows of wire cages. They do not do anything except eat, drink and lay eggs. They cannot run about or go anywhere. Sometimes two or even three birds are kept in one cage. A few people can look after thousands of hens, and eggs can be produced cheaply.

In modern battery systems, the food is supplied by a moving conveyor belt, and the droppings from under the cages can be removed mechanically. The eggs roll out down the sloping floor of the cages so that they are easy to collect.

You may think that this is a cruel and unnatural way to keep hens. A lot of people would agree, but the trouble is that the same people also like to buy cheap eggs! What do you think we should do about this?

Fig. 13.9 *Laying hens kept in battery cages.*

Song of the Battery Hen

We can't grumble about the accommodation.
We have a new concrete floor that's
Always painted white, and a sheet iron roof
The rain drums on. A fan blows warm air
Beneath our feet to disperse the smell
Of chicken manure, and dull days
Fluorescent lighting sees us.
You can tell me if you come by the north door.
I am in the twelfth pen on the left hand side
Of the third row from the floor; and in that pen
I am usually the middle one of three.
But even without directions you'd
Discover me. I have the same orange-
Red comb, yellow beak and auburn
Feathers, but as the door opens and you
Hear above the electric fan a kind of
One word wail, I am the one who sounds loudest
   in my head.

Listen. Outside this house there's an
Orchard with small moss green apple
Trees; Beyond that, two fields of
Cabbages; then, on the far side of
The road, a broiler house. Listen.
One cockerel crows out there, as
Tall and proud as the first hour of sun.
Sometimes I stop calling with the others
To listen, and wonder if he hears me.

The next time you come back here, look for me.
Notice the way I sound inside my head.
God made us all differently,
And blessed us with this expensive home.

Edwin Brock

The two intensive methods of poultry keeping, like the others, have good and bad points:

Good points	Bad points
1 Complete control and full supervision of the birds are possible.	1 There can be risks of disease among overcrowded flocks.
2 Food can be balanced and rationed to meet the needs of the flock.	2 Strict cleaning and sanitation is necessary.
3 Good health care is possible.	3 Costs for food and housing are high.
4 Birds that are not doing well can be culled (removed) from the flock.	4 Some people think keeping hens in battery cages is unnatural and cruel.
5 High levels of production are possible.	

Looking at nearby poultry enterprises

1 Go to several poultry supply stores. Find out what breeds they have on sale.

2 Why do you think there are different breeds of poultry in your area?

3 Talk to anyone in your district who rears poultry. You could either ask your teacher to arrange a visit to a poultry farm, or ask a poultry farmer to visit your school to talk to you about what he does.

4 Find out from these people:
   - the production system they employ – is it free range, semi-intensive, or intensive?
   - why they use that system
   - what breed of birds they keep
   - the space required by the birds (number of birds per square metre) in the poultry house
   - the type of feed used
   - the feeding system and the equipment needed for feeding.

Broilers

PRACTICAL WORK
Looking after a batch of broiler chickens

You will want your broilers to give a quick return on the money you spend on keeping them, so try to raise and sell a batch of broilers within ten weeks at the most. The **broiler hybrids**, *(specially bred for quick growth), are better at producing a quick return than the general purpose pure breeds.*

1 Preparing the broiler house

A house for broilers does not have to be expensive or built by an expert builder. If you want to you can use local materials, like bush poles, and build the house with mud walls and a thatched roof. Later on, when you can afford it, you could build something stronger with cement blocks. The birds will not know the difference. At the end of this chapter, on page 111, you will find a method of building a simple broiler house.

Only one building is needed. The birds stay in the same house from the time they arrive, as day-old chicks, until they are ready for the market.

Two weeks before introducing new birds:

- *the poultry house should be cleaned of all litter from a previous flock*
- *the pen, including the wire mesh and all the equipment, should be washed and disinfected*
- *the equipment should also be tested*
- *fresh litter should then be placed on the floor and treated to prevent any pests and disease infection.*

Broilers can be bought, as day-old chicks, from hatcheries. The chicks should be ordered at least three weeks before the day you want them to arrive.

2 Equipment needed for the broiler house

You will need:

- *a brooder, for rearing the young chicks*
- *small feeders and waterers for day-old chicks*

Fig. 13.10 Equipment for broilers.

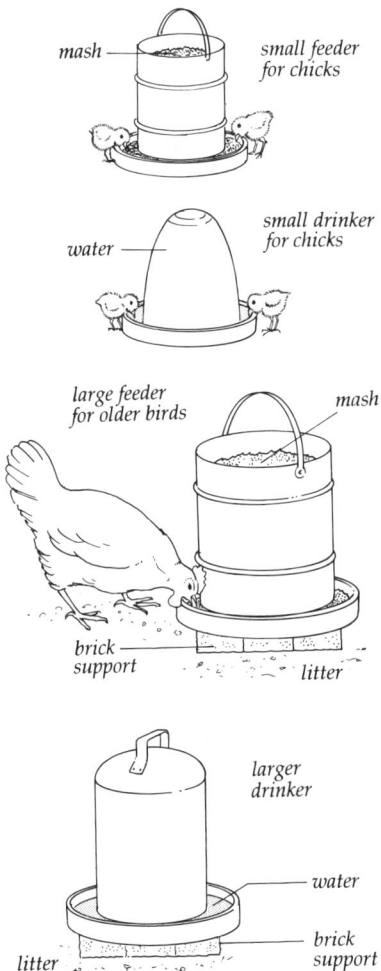

mash — small feeder for chicks

water — small drinker for chicks

large feeder for older birds — mash

brick support — litter

larger drinker

water

litter — brick support

a) Feeders and drinkers.

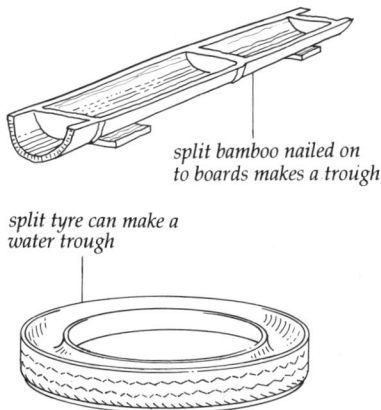

split bamboo nailed on to boards makes a trough

split tyre can make a water trough

- bigger feeders and waterers for older birds
- a supply of litter material, (dry grass, straw or wood shavings)
- broiler mash, a suitable food for broilers
- medicines.

3 Preparing for the arrival of day-old chicks

You have to be ready to receive the chicks when they come! You must:

- **Clean the broiler house.** Make sure that the broiler house is completely clean. Remove all the litter left after other birds have used it. Wash out the house with water and let it dry. Spray the walls and the floor with disinfectant and let it dry again. This is to prevent any pests or diseases being left behind to attack the young chicks.

- **Buy in broiler mash.** Make sure that you have enough to last until the birds are ready for sale. Each bird will require about 4 kg of food during its 8 to 10 weeks of life. Allow a little more in case some birds are slow to reach slaughter weight.

- **Fix up the brooder** and put it in position. Place a thin layer of new litter on the floor. Make sure that the brooder is working well. Check the temperature – with a thermometer if you have one. It should be 35 degrees Celsius at floor level. This temperature will be reduced by 1.5 degrees Celsius each week until the third week.

- **Spread a sheet of newspaper** on the floor of the brooder, and scatter some broiler mash on it just before you put the chicks in. This helps them to start eating as soon as they are ready.

- **Check feeders and drinkers.** You will need small ones for the day-old chicks and larger ones for the older birds. Place a feeder and a drinker where the chicks can find them easily. They should begin to eat and drink on the third or fourth day after hatching. They sometimes do not know where to drink at first. They have never done it before! You can hold them one at a time and dip their beaks in the drinker to teach them where the water is.

4 Brooding

Brooding means caring for the young chicks by keeping them warm and safe when they are small. A mother hen is the best brooder! She looks after her chicks, helps them to find food and water, and covers them with her feathers to keep them warm. But there are not enough mother hens to go round. So we need another way of looking after the young broilers.

A man-made brooder gives the chicks warmth and protection, like a mother hen. A brooder can be bought from a poultry supplier, but it might be expensive. You can make one cheaply that works just as well.

At the end of this chapter two different methods of making a brooder are described, (see page 112). A small area of the floor of the broiler house is partitioned off and covered with newspapers. This area is kept warm with an electric heating lamp positioned above it. The lamp is adjusted to provide a temperature of about 35 degrees Celsius at floor level.

5 Vaccination and debeaking

Your day-old chicks may have been vaccinated at the hatchery against Marek's disease. However, within one week they should be vaccinated against Newcastle disease and Infectious Bronchitis. These two vaccines are usually administered as a combined dose. This is simple to do. Hold each chick gently, and let just one drop fall into one of its eyes.

At the same time Fowl Pox vaccine can be injected into the wing web of the bird. Care should be taken not to puncture any blood vessels. The vaccines are supplied in 1000- and 5000-dose bottles, so if you are buying just a few birds, you could request the supplier to give them these vaccines, and pay a small fee for this service.

*When you are handling the birds, you can **debeak** them. This means removing one third of the upper mandible (beak) to prevent cannibalism. Birds kept close together sometimes tend to peck out each other's feathers, and if a bird begins to bleed, the others may keep pecking it and even kill it.*

6 Feeding and watering

Day-old chicks remain without food for up to 48 hours after hatching. This is because they continue to feed on the remains of the egg yolk within their body. Chicks should be introduced to food as early as possible so that their natural instinct for pecking can develop and they can learn to eat.

Spread sheets of newspaper on the litter inside the brooder and scatter food thinly over it. This helps the chicks to recognize their food, and later to discriminate between food material and litter material which may be similar in colour and texture.

After the first two days, the food can be put into flat trough feeders.

Broiler feeds. It is not easy to mix your own broiler mash to get the right balance of nutrients, including the vitamins and minerals that the birds need. They grow so fast that you cannot afford to make a mistake with their nutrition. It is best to buy broiler mash from a reliable supplier.

During weeks 1 to 5 chicks are fed 'Broiler Starter' which contains higher percentages of protein and calcium for muscle and bone development. In weeks 6 to 10 the birds are fed with 'Broiler Finisher'.

Changing poultry feeds. The change from one feed to another should be made gradually over a week to let the birds become used to the taste, smell and texture of the new feed. Then they will not reject it or become anxious about their food. If that were to happen, their growth rate could be slowed down.

For broilers, make this change in week 4, as follows:

Day	Starter(%) +	Finisher (%) =	Mix
1–2	75	25	= 75:25
2–4	50	50	= 50:50
5–6	25	75	= 25:75
7–>	0	100	= 100% finisher

Another question is, 'How much food should broilers be given?' The answer is, 'They should have as much as *they want!' This is referred to as 'Ad lib' feeding. It does not pay to ration their food. We want them to grow very fast. The table in Fig. 13.11 shows how much you can expect them to eat as they grow older and heavier.*

Age in weeks	Weight of broiler (grams)	Food eaten per day (grams)
0	40	10
1	100	30
2	230	50
3	410	60
4	650	70
5	940	80
6	1200	90
7	1500	100
8	1800	110
9	2000	120
10	2300	130

Fig. 13.11 *How much food broilers eat.*

The graph in Fig. 13.12 shows how the rate of growth changes as the birds grow older. The steeper the curve, the faster the rate of growth.

*Broiler chicks are specially bred for rapid growth, but without the right feeding, they cannot keep up the high rate of **live weight gain** expected of them.*

Water. *Broilers should have a constant supply of clean water. The broiler mash that they eat is dry, so they cannot eat it without drinking water. Try not to spill water on the litter. Wet, cold litter does not decompose properly.*

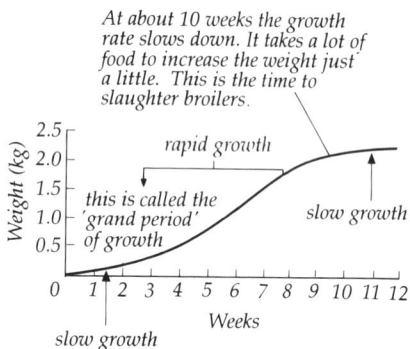

Fig. 13.12 *A growth curve for broilers.*

Layers

1   Chickens kept for laying eggs must be allowed to mature to five or six months of age, (20–22 weeks). Then they are called **point of lay pullets**. You may be able to buy point of lay pullets from a poultry farm,

but there is really no reason why you should not rear them yourself, just as you would rear broilers. It just means that you have to wait longer for a return on the money spent on the birds and their feed.

2 When the birds begin to lay make sure they have:

- comfortable nest boxes lined with fine dry grass.
- a continuous supply of layers' mash – a balanced feed for laying birds.
- a continuous supply of fresh water. They need a lot of water to produce eggs.
- you may also have to supplement the diet with crushed oyster shells or limestone grit. These are both sources of calcium needed to prevent your hens laying eggs with thin shells.

3 The following routines and procedures described for caring for the birds and the litter are the same for layers as for broilers, and should not prove difficult.

Routines for looking after poultry

Chickens are nervous birds, so try to keep them calm and quiet. Remember they are much smaller and weaker than you, and you could frighten or hurt them without meaning to. You can:

1 *Let them know you are coming.* Never go into the poultry house suddenly! Make a little noise, or say something, as you come near to the house. Then they will know you are there.

2 *Move slowly and quietly* when you are in the poultry house. Sudden movements frighten them. Do not throw paper sacks about when you are filling the feeders, or throw grass about when you are putting litter on the floor.

3 *Follow a definite routine each time,* so that the birds learn what to expect. Let them think you are their friend who brings them food, not their enemy who is going to catch them and eat them! Perhaps you *are* going to eat them, but they don't know that yet.

Daily routines:

1 *Feed them first.* That is what they want. When they are eating happily you can be doing other

things that need to be done. Check the feeders and fill them up as necessary. They give a steady supply of fresh food.

If you are using home-made troughs, make sure that any old food is eaten before adding more to the trough. It is a mistake to put new food on top of old food, because the old food may stay there underneath and go mouldy. This could be very bad for your birds.

2 *Check the drinkers.* Make sure that they are clean. If they are dirty, take them outside and wash out dirt and droppings. This is most important because diseases are spread by drinking dirty water. Fill the drinkers and take them back in.

3 *Collect eggs* from the layers, and as you do this, watch out for **broody hens**. Sometimes laying hens sit on the eggs and make a threatening noise when you come near, while fluffing up their feathers to look as fierce as they can. They are trying to hatch their eggs, which is only natural, but broody birds are no longer laying. They should be put into a cage with a wire floor for a few days. This cools them down so that they forget about the warm nest. Then they should come back into lay. If they persist in being broody, they must be culled from the flock.

Keep a careful record of the number of eggs laid each day.

4 *Look around carefully* to see if all the birds look healthy and are moving about, and eating and drinking normally. If you see any that look sick, tell your teacher.

Weekly routines

1 Once a week check to see if the litter has become compacted, (hard). If it has, dig it over with a garden fork and rake it smooth. This lets air into the litter and helps it to break down. As you dig, move slowly and quietly so as not to disturb the birds.

2 Add fresh litter. Carry it low down and spread it gently.

Fig. 13.13 *Raking over the litter.*

Fig. 13.14 *Catching and weighing a broiler.*

a) How to catch a bird with a catching hook.

b) How to hold a broiler.

wrap a cloth fairly tightly round the broiler to hold it still

c) How to weigh a broiler.

3 Keep a record of the live weight gain of your broilers. To do this, catch five birds and weigh them. Write down the weight of each one and work out the average weight per bird. If you do this every week, you can see how fast they are growing, and plot a **growth curve** like the one shown in Fig. 13.12 on page 102. When catching birds, keep your hand low and take hold of their legs. Never chase them round the broiler house. Another way is to make a catching hook of strong fencing wire. The hook just fits over the thin part of a chicken's leg. A good time to catch the birds is when they are busy eating.

Cleaning out the poultry house between batches

After each batch of birds is sold off, it is very important to *clean out and disinfect* the poultry house. Remember to remove all the deep litter to the compost heap, wash the house out thoroughly and spray it with disinfectant (Fig. 13.15). Then no pest or disease organisms will be left behind to infect the next batch of young chicks coming in.

The rule is, 'all in, all out'. Never be tempted to leave the old litter in the house and just bring in another batch of chicks. That would be asking for trouble.

Disease prevention and control

Parasites of chickens

A parasite is a pest that lives on another living organism. The table in Fig. 13.16 shows two parasites that can affect chickens. Note the symptoms, (signs), of attack, and the methods of control.

Parasites	Symptoms	Method of control
1 Lice	Small grey insects moving over the skin and on the feathers. Chickens loose blood, and become weak.	Malathion 5% insecticide dust.

Fig. 13.15 *Clearing out the deep litter.*

| 2 | Red mite | Tiny red eight-legged mites living in cracks in the walls, and feeding on birds at night. Chickens loose blood, and become weak. | Malathion 5% insecticide dust. |

Fig. 13.16 *Parasites of chickens.*

Diseases of chickens

The table in Fig. 13.17 shows two diseases that can affect chickens. Note the symptoms, (signs), of attack, and the methods of control.

Diseases and causal agents	Symptoms	Method of control
1 Fowl typhoid (bacterium)	Coughing, sneezing, discharge from the nostrils and eyes. Noisy breathing. Green, smelly droppings.	Furazolidone mixed in food. Vaccination. Good hatchery sanitation.
2 Fowl pox (virus)	A cheesy yellowish membrane in the mouth and throat. Breathing difficult. Smell from mouth is always foul.	Vaccination of non-infected birds.
3 Coccidiosis (protozoon)	Diarrhoea. Blood in the droppings. Anaemia. Ruffled feathers. Sudden deaths of young chicks.	Sulphamezathine added to the drinking water. Keep litter dry.
4 Marek's disease (virus)	Paralysis. The birds' legs are stretched out stiffly.	No cure. Vaccination can prevent it.
5 Newcastle disease (virus)	Difficulty in breathing. Coughing. Sneezing. Paralysis and death follow. Greenish, white foul smelling droppings.	Vaccination can prevent it.

6 Pullorum (bacterium)	Septicemia (blood poisoning). Difficult breathing. White foaming droppings. Loss of appetite.	Cull affected birds. Clean and disinfect the poultry house Fumigate incubator and eggs.
7 Coryza (bacterium)	Coughing. Sneezing. Yellow discharge around the eyes.	Treat with antibiotics in drinking water.
8 Infectious bronchitis (virus)	Coughing. Sneezing. Discharge from the nostrils. Watery eyes.	Vaccination. Birds that recover have immunity.

Fig. 13.17 *Diseases of chickens.*

Fig. 13.18 *Candling eggs to see if they are fertile.*

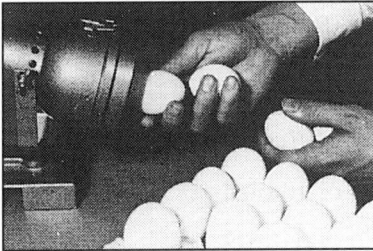

a) A candling machine in use at a hatchery.

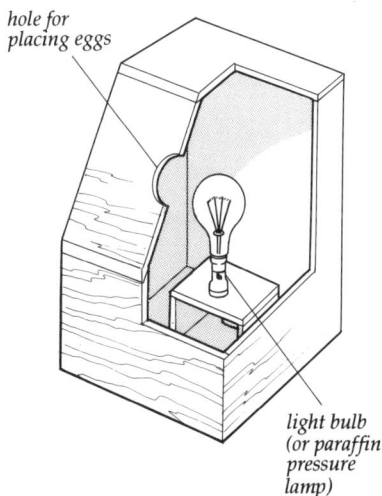

hole for placing eggs

light bulb (or paraffin pressure lamp)

b) A home-made candling machine.

If you notice any signs of such pests or diseases, tell your teacher straightaway. If any birds die, keep a record in your diary of when they died and the cause of death.

Natural brooding and Artificial incubation

It is fascinating to watch a mother hen incubating and hatching a clutch of eggs. You can arrange this quite easily.

1 Obtain 10–12 fertile eggs, either from a hatchery, or from a flock of hens that have a cockerel running with them. Only hens that have mated with a cockerel can produce fertile eggs. It is possible to check if eggs are fertile by shining a strong light from behind them. This is called **candling**. If an egg is fertile, a shadow is cast by the small embryo inside.

2 Let a **broody hen** sit on the eggs and incubate them until they hatch. To see if a hen is broody, (ready to sit on eggs), place one egg in front of her when she is on the nest. If she pulls it under her with her beak and makes satisfied clucking noises, she is broody.

3 Keep the hen supplied with food and water. She may not come off the nest much at all for 21 days. This is the **incubation period** when the young chick embryos are growing inside the eggs.

4 After 21 days the chicks should hatch out. Watch them hatching if you can. Then watch how the

mother hen leads her young chicks about in search of food, and how she takes them under her wings for warmth and protection.

Records and marketing

Keeping records

Keep a record of **costs** and **returns** for broiler production, as you would for any other farm enterprise. An example of an account sheet is shown in Fig. 13.19. Copy it into your exercise book, but fill in the costs and returns of your *own* enterprise. Each time you buy or sell anything, write it down in the appropriate place.

You can also keep records of the live weight gain of your broilers, and the egg production of your layers.

	COSTS				RETURNS	
Date	*Item*	*Amount*		*Date*	*Item*	*Amount*
	45 day old chicks				Sale of 20 birds	
	Transport for chicks				Sale of 10 birds	
	Transport for litter				Sale of 10 birds	
	Broiler starter 100 kg				Sale of 5 birds	
	Broiler finisher 150 kg					
	Antibiotics					
	Electricity					
	Building depreciation or rent					
	TOTAL COST				TOTAL RETURN	

TOTAL RETURN	
TOTAL COSTS	
PROFIT	

Fig. 13.19 A Record Sheet for broiler production.

Marketing of broilers

1 *When should you sell?* The best time to sell is *after* the broilers have reached a good size and weight, but *before* they have begun to cost too much to feed.

2 *How to tell when the birds are ready.* They should be ready at about 8–10 weeks old and weigh about 2–2.5 kg. If you have been keeping a record of the live weight gain of your broilers this will help you to understand why you should sell them at this time. Your graph, like the one in Fig. 13.12, should show that the rate of growth is steady up to between 6 and 8 weeks. Then it slows down and the graph becomes less steep. After this it will take a *lot* of food to get a *small* increase in weight. If you keep the birds too long, they will cost money to feed, but not give any extra profit. This is the time to sell them.

3 *Selling live broilers.* If you sell live birds, you know their meat will stay fresh until your customers are ready to eat them. Sometimes this is what the customer wants, and it is the easiest way for you to sell them. You do not have to kill, pluck and clean them.

4 *Selling broilers ready for cooking.* Sometimes people like to buy their broiler ready for cooking. Then you have to kill, pluck and clean it yourself. This must all be done quickly, and the bird must be sold, before the meat has time to go bad.

Marketing of eggs

1 Try to keep the eggs clean. This means removing droppings from the nest boxes, and putting in fresh, dry grass as nesting material.

2 If any eggs do get dirty, clean them with a damp cloth or with some steel wool. It is not a good idea to wash the eggs in water, as some of the dirt might penetrate the porous shell.

3 Keep the eggs in proper trays or eggs boxes, to prevent them from breaking, and sell them while they are fresh.

4 Try to keep up a good reputation with your customers for clean fresh eggs. If any eggs are a funny shape, keep these ones for eating or cooking.

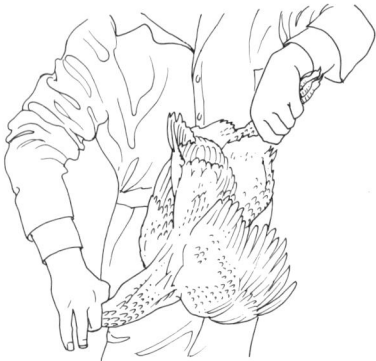

1) killing the bird

2) plucking off the feathers

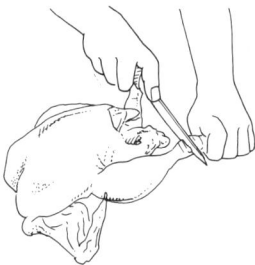

3) cutting off the legs

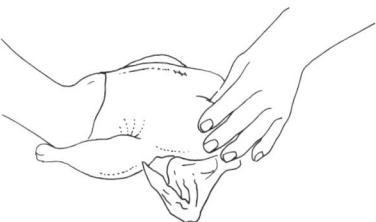

4) drawing the bird

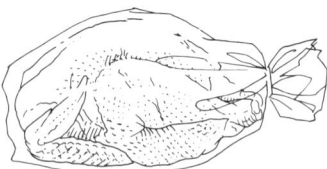

5) packing

Fig. 13.20 *Dressing a chicken.*

How to 'dress' a chicken

To learn how to do this it is best to watch an experienced poultry keeper do it. The method is not hard to follow. Stop feeding the bird 24 hours before it is due to be killed. This will make the cleaning easier The food in its intestine will pass through in that time.

1   Kill the bird by breaking its neck, and cut off the head. Let the blood drain from it by holding it upside down.

2   Pluck off all the feathers while the bird is still warm. If it gets cold, the feathers will be harder to pull out. You might have to dip the carcass in very hot water before plucking to make the feathers come out more easily.

3   Cut off the legs.

4   Draw the bird. Carefully cut with a sharp knife all round the vent, under the tail. Do not cut the rectum, or the droppings will make the carcass dirty. Pull out all the internal organs. The heart, liver and gizzard are good to eat, and they are usually put back inside and sold with the broiler. The crop can be pulled out from the neck end.

5   Put the carcass in a clear plastic bag to keep it clean and to keep flies off it.

Is it worthwhile to keep chickens?

1   After you have sold your broilers, discuss these questions with your classmates and try to answer them:

   – was it a worthwhile project?

   – did you make a profit?

   – how much?

   – would you do it again?

   – how could you tell from your records whether or not you made a profit?

   – is there a good market for broilers in your area?

Fig. 13.21 *Building a broiler house.*

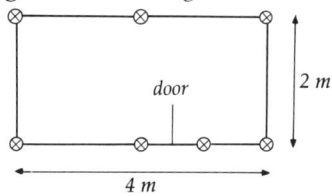

a) A ground plan

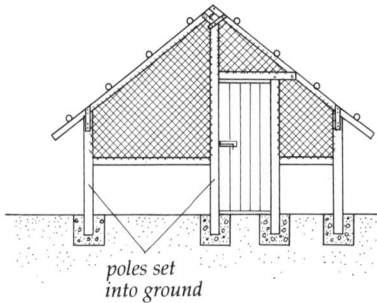

poles set into ground

b) Front view

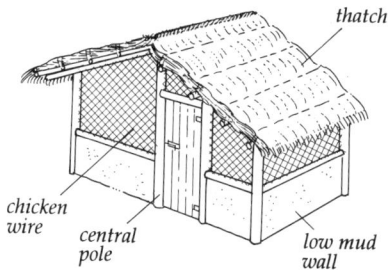

thatch

chicken wire

central pole

low mud wall

c) The completed house

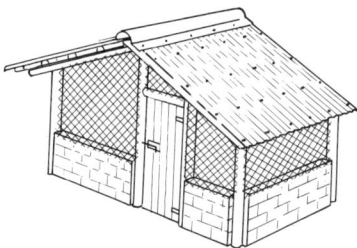

d) The same type of building made with block walls and corrugated iron roof

– do people like to buy them?
– could you keep more next time?
– could you keep broilers at home?

2 If you are keeping laying hens, do the same exercise for them.

3 When you have agreed on the answers, write them down in your notebooks.

How to build a poultry house

1 Draw a plan

Work out what size to make the house, and draw a plan to scale. The smallest house worth building, for broilers, would be 3 m × 3 m square, but you can make it bigger if you want to. Align the house 'east-to-west'. Do you remember why? (See page 93.)

As a guide, one square metre of floor space is needed for every *nine birds* when they are young. So first work out how many broilers you want to keep, and how many you think you can sell. Then make the house big enough for this number. An example may help: a small house of only 9 square metres in floor area (measuring 3 m × 3 m) will hold 9 × 9 = 81 broilers when you first put them in. As they reach slaughter weight, they will be thinned out and the remaining birds should still have just enough floor space.

Layers, on the other hand, should be allowed one square metre for every three birds, so you have to think in terms of a bigger house for them.

2 Decide what materials to use

- The main structure of the house is best built of creosoted poles set into the ground. The poles support the roof.
- The walls of the house shown in Fig. 13.21d are built of concrete blocks, about 1 m high.
- The house in Fig.13.21d is open and well ventilated to let in light and air. At the same time its roof and walls protect the birds from sun and wind. The roof is made of corrugated iron sheets. The floor is made of concrete.

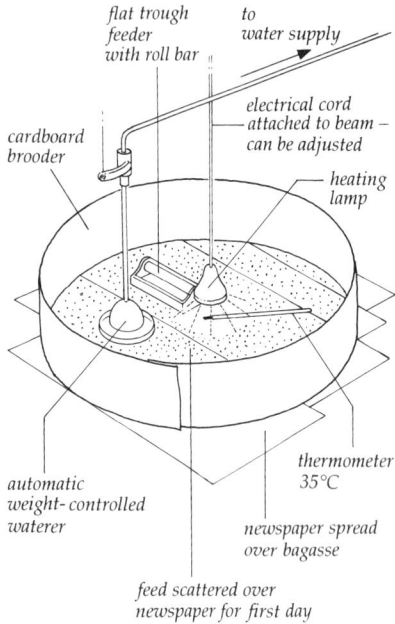

Fig. 13.22 An electric lamp brooder.

Fig. 13.23 A brooder made from an old oil drum.

- If the cost of this house seems too great, you could make a house using local materials, like bush poles, with mud walls and floor, and a thatched roof (see Fig.13.21c). Instead of wire mesh you could use thin bush poles nailed 5 cm apart. These materials will not last so long, but they are cheap, and good enough to start with.

3 Fit a strong door

Also a good lock!

Making a brooder

Two types of home-made brooder are shown in Figs. 13.22 and 13.23.

Electric lamp brooder

1 This brooder has an electric lamp that gives out heat, hanging about 60–80 cm above the floor, to keep the chicks warm.

2 The lamp is raised or lowered until the right temperature is reached. This should be 35 degrees Celsius, (just below blood heat), at floor level, under the lamp. This should feel warm to your hand.

3 A wall of cardboard or hardboard is built to prevent the chicks going too far away from the heat or getting lost.

4 A feeder and a drinker are placed close to the lamp where the chicks can find them.

Oil drum brooder

1 This brooder is made from an old oil drum cut in half and hung from the roof.

2 The drum has two strong wire loops, made of fencing wire, and ropes are passed through these loops to hang it up.

3 The ropes can be shortened or let out, to raise or lower the brooder.

4 A paraffin oil lamp is put inside to keep the chicks warm.

5 You will have to experiment with the lamp, putting it inside and raising or lowering the drum, until the right temperature is reached. This should be 35 degrees Celsius, at floor level, close to the lamp.

6 A feeder and a drinker are placed near to the drum so that the young chicks can find them easily.

Things to do

1 Make a drawing of the type of brooder you are using at your school.

2 Label it to show the different parts and what they do.

Questions

1 (a) What is meant by a 'dual purpose' breed of poultry?

   (b) Give an example of one such breed.

   (c) What are the advantages of keeping dual purpose birds?

2 (a) List four features of a good broiler chicken.

   (b) When a batch of broilers has been sold off, how would you prepare the broiler house to receive a new batch?

3 (a) Draw a brooder for day-old chicks, and label the parts.

   (b) Explain how to prepare the brooder for the arrival of day-old broiler chicks.

   (c) How would you supply the chicks with food and water?

4 (a) What is 'deep litter'?

   (b) Describe the daily routines for looking after laying hens on deep litter.

   (c) What could go wrong with the litter? How can you help to keep it in good condition?

5 (a) Name one external parasite of chickens.

   (b) Describe the parasite and the harm it causes to the birds.

   (c) How can it be controlled?

14

AGRICULTURE AS A BUSINESS

When you have read this chapter you will be able to:

- *explain what is meant by a **business**.*
- *describe the resources needed by all businesses.*
- *describe the resources needed for a farm business.*
- *explain how businesses make money.*
- *plan a farm business enterprise.*
- *describe the different types of agricultural businesses:*
     - *primary production*
     - *marketing*
     - *manufacturing.*
- *describe how some local agriculture businesses work.*

Business is about buying and selling. There are many different things that can be bought and sold, and so there are many different kinds of business. People go into business to make money, and basically they all do the same things:

- They supply something that people want or need.
- They charge a price that their customers are willing to pay.
- If they can get more money from selling goods or services than it costs to provide them, their business makes money.

Here are some examples of small businesses.

Fig. 14.1 *Different businesses.*

a) Dressmaking

b) Fixing bicycles

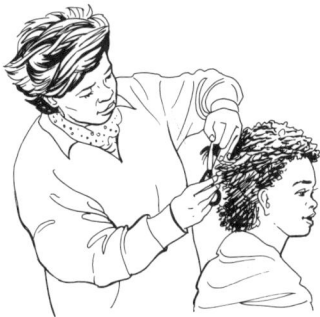

c) Hairdressing

d) Shopkeeping

- A lady who is a good dressmaker might decide to buy some cloth, and make clothes for sale on her sewing machine. She can sell them to people who are not so good at dressmaking.

- A man who is clever at fixing bicycles could fix them for people who do not know how to fix their own bicycles. They pay him to do it for them.

- A hairdresser can create nice hairstyles for ladies. They pay her to do this for them, because they cannot create such good hairstyles themselves.

- A shopkeeper buys goods from different suppliers, and sells them at a slightly increased price. People pay the shopkeeper for collecting and stocking all the things they need.

- A higgler, who goes around selling vegetables, makes a profit by charging a little more than the cost price. She helps people by saving them the trouble of going to the market, and they are prepared to pay something for this service.

This is fair and reasonable. No one is being forced to buy anything, and the customers are getting what they want at a price they can afford. Everybody is happy.

Resources of a business

Before you can run a business, you must have certain things in place to help you start. These necessary things are called the **resources** for the business. For example, you need:

- **A place to work** – a house, shop, factory or farm, where you can carry on the work and where your customers can come.

- **Materials** to work on.

- **Knowledge and skill** – you have to be good at what you do, so that people will want to buy what you produce. You also have to know how to work out the cost of your materials, and the prices you can charge your customers – not too high, or people will not buy, and not too low, or you will not make enough money.

- **Capital** – this means the money to buy the stock and the tools you need, and also to cover running costs.

Look again at the pictures of people running their small businesses in Fig.14.1. What 'resources' does each of them need?

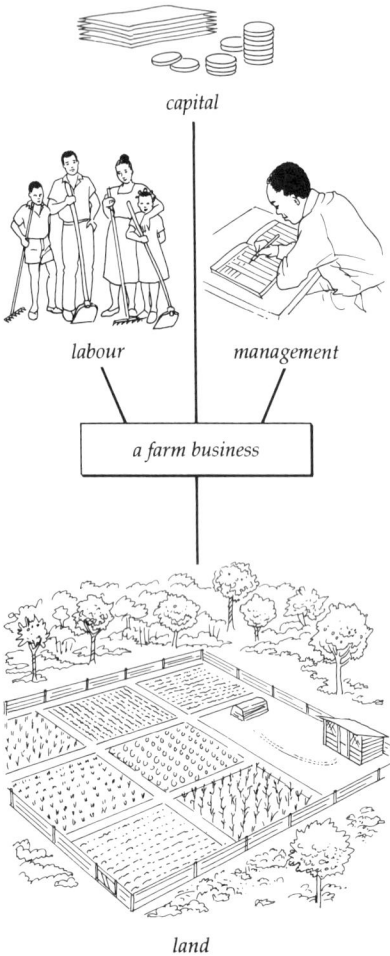

Fig. 14.2 *The resources needed for a farm business: land, labour, capital and management.*

Resources for the agriculture business

Agriculture is a business, like any other business, and farmers do what all other business people do. They use their resources, and their skill, to produce things that other people want to buy.

Think of how you would set about running a farm business. The first thing you would do is check the resources you have to start a farm and keep it going. You would need:

- **Land**. You need somewhere to grow your crops and keep your animals. If you have a lot of arable land, you might decide to grow crops for sale. If you have just a small piece of land, you might keep chickens or rabbits.

- **Capital**. This means enough money to pay for all the **inputs** you would need to get the farm started – such things as seeds, fertilizers, animal houses, tools, equipment and labour. You would need enough money to keep going until you sell some produce and begin to get some money back again. Farmers can use **capital** from their own savings, or they may be able to get **credit**, (a loan from the bank). They can 'start small and build up'. This means putting some of the profits back into the business to make it grow.

- **Labour**. Who will do the work? If you have a small farm, it can be run by your family. Bigger farms need to employ people to work in the fields.

- **Management**. This is an important resource. A farm cannot succeed unless someone has the knowledge and skill to run it. Could you be your own farm manager?

If you understood this, you are beginning to 'think like a farmer'.

Planning to start a farm

Divide the class into groups. Then, as an exercise, let each group try copying and completing a planning chart, as though you were actually going to start each of the five agricultural enterprises. 'Vegetable Production' has been started for you.

	Enterprise 1 Vegetable Production	Enterprise 2 Broiler Chickens	Enterprise 3 Laying Hens	Enterprise 4 Beef Cattle	Enterprise 5 Dairy Cattle
Resources needed	**Land** A fenced garden	**Land**	**Land**	**Land**	**Land**
	Capital for buying tools, seeds irrigation	**Capital**	**Capital**	**Capital**	**Capital**
	Labour our family	**Labour**	**Labour**	**Labour**	**Labour**
	Management myself	**Management**	**Management**	**Management**	**Management**
Produce for sale	peas, beans carrots tomatoes potatoes spinach				

Fig. 14.3 *Primary production - selling farm produce.*

Types of agricultural businesses

There are other businesses in the agriculture industry besides the **primary production** described already, i.e. growing crops and raising animals for sale. There are three main types of agricultural business: production, marketing and processing.

Marketing

Before a farmer can grow anything, he has to buy inputs such as seeds, fertilizers, chemicals, tools and machinery. Agricultural suppliers make a business out of supplying these things. They buy in supplies in bulk at wholesale prices (i.e. cheaply), package them up, and sell them on at retail prices (i.e. more expensively) to farmers. They are providing a useful service, and the difference between the wholesale price and the retail price gives them their profit.

A **marketing agent** buys produce from the farmer, moves it to the market and sells it. Sometimes the market may be in another country. For example, citrus fruits such as oranges and grapefruit are produced in the Caribbean, but they may be sold in Europe.

The marketing business has to have resources to run it, and it has to make money like any other business. The resources are:

- the knowledge and skill of the agent who arranges for the fruit to be marketed abroad
- offices, telephones, fax machines, required to run the business.

The money made by the business comes from the overseas customers, and this makes the business worthwhile.

Both the farmer and the marketing agent make money from their businesses. The farmer sells the produce for an agreed price. Then the marketing agent sells it on to the foreign customer at a *higher* price. Oranges cost more in Europe than they do in the Caribbean Islands.

Processing

Not all farm products are sold in the form in which they leave the farm. A farmer sells sugar cane, but the customer wants sugar. Someone must process the cane to make the sugar. This is done at the factory, and the factory is a business of its own. At a processing factory, the produce of the farm is processed to make sugar for sale.

Many other ways of processing farm produce could be mentioned. They are usually concerned with the four objectives in the following table. See if you can think of an example of each one:

Longevity	– making the product last a long time.
Utility	– improving its ease of use. Making the product more readily available.
Portability	– ease of transport.
Versatility	– allowing it to be used in many ways.

The processing business, like the marketing business, occupies a place between the farmer and the customer who buys the final product.

A manufacturing business has **resources**, and it **makes money**. The money made by the business comes from selling sugar. Again, there has to be a suitable **price structure**, so that:

- farmers receive a fair price for their sugar cane
- customers pay a fair price for the sugar
- and the factory owner makes a living as well.

Local businesses

There will be agricultural businesses near your home or school. They may be sugar factories, copra factories, fruit or vegetable canning factories, fish freezing and canning factories, or other processing activities found in your area.

1 Divide into groups. Let each group try to think of some agricultural businesses in your area and make a list of them. For each business ask:
 - What do they *do* in this business?
 - What goods or services are they selling?
 - What are the resources of the business, (land, machinery, marketing skills, labour)?
 - Who is managing the business?
 - How does the business make money?
 - Who are the customers?

2 Afterwards, let each group report to the class on one of the businesses they have discussed.

3 If you can answer questions like the ones above, you can learn to run a business.

A visit to an agricultural business

You can learn a lot by watching people at work in the world outside school, and listening to them talk.

If possible, visit a local agriculture business, or ask someone from the business to come and tell your class about it. Then you will learn at first hand how the business works.

Use a business questionnaire. Be ready to ask questions and write down the answers in as much detail as you can. This information could be useful to you one day, if you start your own business.

Fig. 14.4 Finding out about an agricultural business.

QUESTIONS

- What does the business *do?* Is it concerned with:
 a) primary production?   b) marketing?
 c) manufacturing?

- Who runs the business?
- What are the resources of the business (land, tools and equipment, labour, skills)?
- Who are the customers?
- What goods or services do the customers receive?
- What needs in the community are met by the business?
- What do the people running the business hope to get out of it?
- Finally, ask if the business life is a good life?
  - what are the good things about it?
  - what are the bad things?

Questions

1 (a) What is a 'business'?

(b) Give one example of a small business. Explain what product or service is being offered to the customer, and how money can be made.

2 If you wanted to grow vegetables for sale, describe:

(a) Four resources needed to start and run your business.

(b) How you could make a profit.

3 (a) What is meant by 'agricultural marketing'?

(b) Explain, giving an example, how a marketing agent can make money.

(c) What is meant by the 'export market'? Give an example to show how money can be made in this market.

4 (a) Name a farm product that is processed before being sold.

(b) Describe how it is processed.

(c) How do the customers benefit from the processing of this product?

5 Describe a visit you have made to an agricultural business, and say:

(a) What work is being done in the business.

(b) What resources are needed to run it.

(c) What is the end product.

(d) Who buys the end product.

15

THE PLANT ENVIRONMENT

When you have read this chapter you will be able to:

- *say what is meant by the term environment.*
- *make a list of environmental factors that affect the growth of plants.*
- *explain the favourable or unfavourable effects of each factor on plants.*
- *say what farmers can do to cope with each factor, and to improve conditions for their crops.*

Fig. 15.1 Imagine you were a plant!

Environment means surroundings. When we talk about the **plant environment** we mean everything surrounding a plant that influences the way in which it grows.

It is important to understand the **environmental factors** influencing plant growth, because they greatly affect crop yields. If we understand the plant environment, we can provide suitable conditions for our crops and help them to grow better. Try to look at it from a plant's point of view.

What is it like to be a plant?

It is hard for us to imagine what it must be like to *be* a plant. Try it! Suppose that instead of being a group of students you were a group of plants standing close together in the garden. This is just a game, but it can help you to understand the problems that plants have.

Think about it:

- You have roots. Half of your body is below the ground.

- You cannot move about.
- You may not like the soil you are growing in. What can you do about it?
- You may not like your neighbours, and they are standing much too close. They are bigger than you. Can you do anything about them?
- You need light, water, nutrients and space for your roots, but other plants compete with you for all these things. Is there anything you can do about the competiton?
- What about the climate and the weather? Can you change them?
- Can you help yourself at all, in any way?

You realize now that:

- plants have no choice about where they live, or who their neighbours are.
- plants are helpless to change much in their own environment.

You must be glad you are a student and not a plant.

Fortunately, *we* are not helpless. As intelligent people, we can make choices about where our crop plants live, and we can do a lot to improve their environment. A plant needs a farmer for a friend. Then it can expect a better life. Farmers are not just being nice to plants, of course. They look after them because they want crops to yield well, and produce food.

Things that can affect plant growth

The main environmental factors affecting the growth of plants can be grouped under the following three headings:

1 Soil factors

2 Climatic factors

3 Biotic factors

The first two groups of factors are non-living, and are called the **abiotic factors**. The last group is made up of living organisms, and these are called the **biotic factors**.

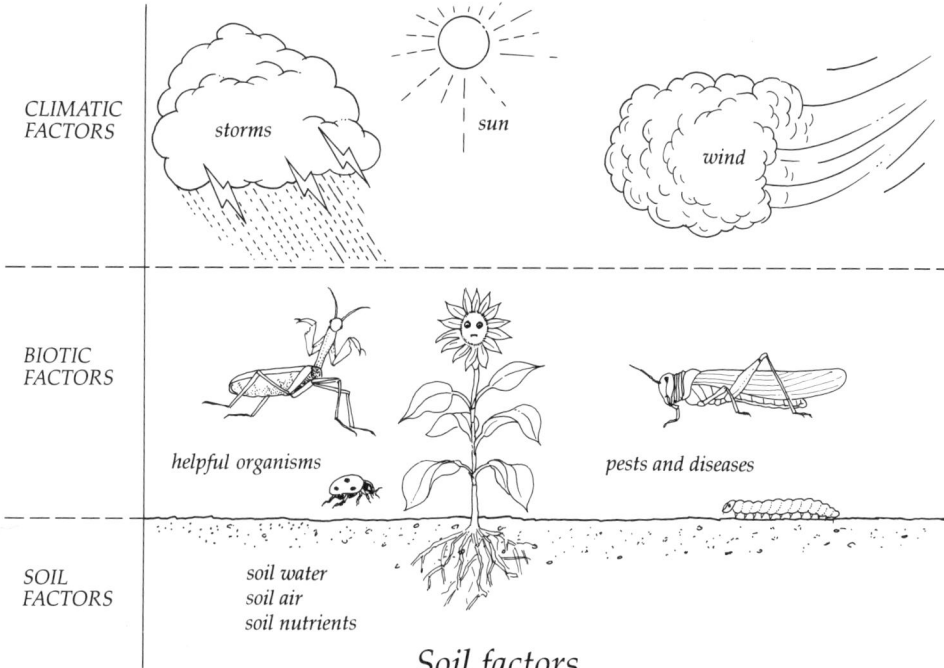

CLIMATIC
FACTORS

storms

sun

wind

BIOTIC
FACTORS

helpful organisms

pests and diseases

SOIL
FACTORS

soil water
soil air
soil nutrients

Fig. 15.2 *Factors affecting plants:
climatic, biotic and soil.*

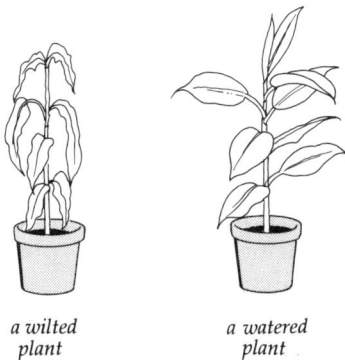

*a wilted
plant*

*a watered
plant*

Fig. 15.3 *A plant needs water.*

Soil factors

The soil is bound to have a major effect on plants. Almost half the body of each plant is under the ground. Roots reach out underground into a large volume of soil, and plants rely on the soil to supply them with water, air, nutrients and support. Think how the following three factors can affect plants.

Water

A plant cannot grow at all without water. Up to 90% of the total body weight of some vegetables is made up of water, and all this water comes from the soil.

Problems	*How we can help*
When the soil lacks water, plants wilt and die. Soil organisms also die.	We can improve the **water holding capacity** of soil by adding compost or animal manure. These rot and form humus, which can hold a lot of water.

Air

Air is present in the soil, between the soil particles. The soil air contains oxygen, which makes it possible for plant roots and soil organisms to respire.

compacted soil:
roots cannot penetrate
little air for respiration

loose, friable soil:
roots spread easily
air can enter the soil

Fig. 15.4 *A plant needs air.*

this plant is
hungry

this plant is
well fed

Fig. 15.5 *A plant needs nutrients from the soil.*

Problems	How we can help
1 When soil is short of air, (e.g. when it is tightly compacted, or waterlogged), plant roots and soil organisms suffer from lack of oxygen, and eventually die.	1 Improve 'aeration' by digging or ploughing. 2 Cultivation breaks up the soil. It allows water and air to penetrate more easily.

Nutrients

Certain chemical salts in the soil are needed as nutrients by plants. Plants 'feed' by taking in these nutrients dissolved in water.

Problems	How we can help
When soils lack nutrients, plants suffer from **nutrient deficiency** and growth is poor.	We can correct nutrient deficiency by putting manure, compost or fertilizers on to the soil.

Climatic factors

Plants are exposed constantly to the weather. They have to cope with **temperature changes** – heat by day and cold by night. Variations in temperature and **rainfall** from one season to another are beyond their control. Climatic factors have an enormous effect on plants. They are discussed more fully in Chapter 17.

Temperature

Plants will not grow well if the temperature is either too hot or too cold. Each species of plant has a range of temperatures within which it can survive happily. This is why some species grow best in warm climates and others in cool climates.

For each species there is:

- a **minimum** temperature below which it will not grow.
- a **maximum** temperature above which it will not grow.
- an **optimum** temperature at which it grows best.

Fig. 15.6 *Compare these two photographs. What do they tell you about the climatic conditions in each place?*

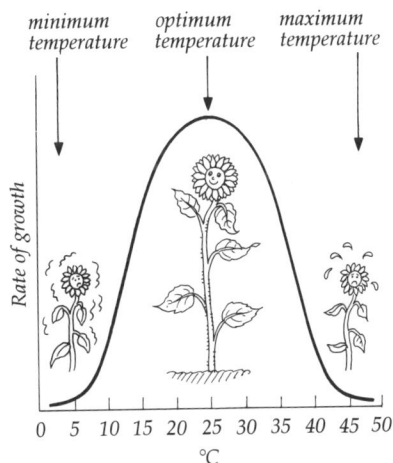

Fig. 15.7 *A plant needs a suitable temperature.*

Problems	How we can help
1 Plants will not grow if the temperature is too hot or too cold.	1 We can plant at the right time of year. 2 We can give tender plants shade when it is too hot.

Rainfall

Some plants require more rainfall than others, and different places have different amounts of rain.

Problems	How we can help
1 Plants will not grow well if the rainfall is too little or too much.	1 We can find out how much rainfall to expect during the year. 2 We can choose plants that are suited to the rainfall of our area.

Other climatic factors

Some places are subject to exceptional events, like hurricane-force winds, or long periods of drought.

Problems	How we can help
1 These freak conditions can do great harm to crops.	1 We can find out what to expect, and grow crops that can survive freak conditions. 2 We can introduce windbreaks, or irrigation systems.

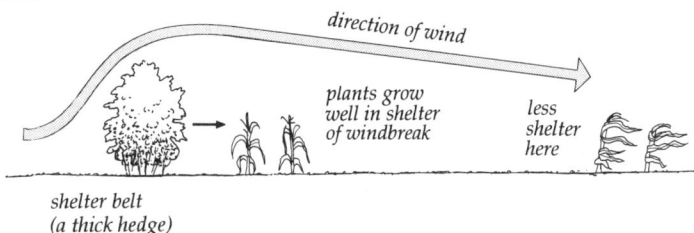

Fig. 15.8 *Protecting plants against wind.*

Fig. 15.9 *Some organisms that live in the soil.*

a) *Earthworm*

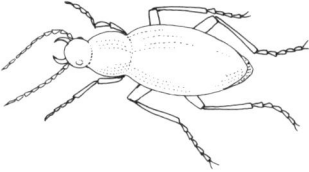

b) *Ground beetle*

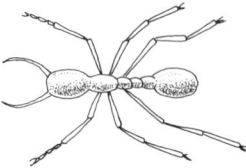

c) *Ant*

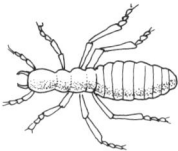

d) *Termite*

Fig. 15.10 *Plants need to be free of competition from weeds. Which plant is suffering from weed competition?*

Biotic factors

Other living organisms in the environment affect the plant. There are many of them, such as micro-organisms and small animals in the soil, other plants and weeds, birds, animals, and human beings.

Soil organisms

Micro-organisms, like bacteria and fungi, are present in the soil in large numbers. They feed on dead plants and animal waste, causing them to decompose. The products of decay are nutrients needed by living plants. Soil organisms have an essential part to play in recycling plant nutrients.

Problems	*How we can help*
1 The soil organisms may be few in number due to lack of organic waste to feed them.	1 We can add compost or manure to the soil.
2 Therefore recycling of nutrients cannot take place, and the soil is not fertile.	2 This will encourage the soil organisms to grow.
	3 Nutrients will be recycled, and the soil will become more fertile.

Larger organisms

Examples of these are earthworms, beetles, ants and termites which burrow into the soil. These are helpful organisms on the whole, because they improve drainage and aeration.

Problems	*How we can help*
When the soil has few organisms, air and water cannot penetrate so easily.	We can add compost and manure to encourage small soil animals to increase their numbers.

Fig. 15.11 Helpful and harmful insects.

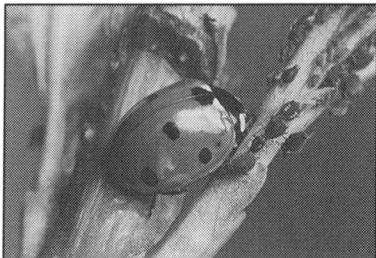

a) Two helpful insects . . . bee and ladybird

b) Two harmful insects . . . locust and aphid

Other plants and weeds

Competition from neighbouring plants can have a big effect on plants.

Problems	How we can help
1 Nutrients, light and water are used up by neighbouring plants.	1 We can make sure that crop plants are spaced correctly in rows, so that they have enough room to grow.
2 Crop plants can suffer.	2 We can keep weeds down, so that the crops have no competition.

Insects

There are huge numbers of insects of all kinds sharing the environment with plants. Some are helpful, like bees that pollinate flowers, or ladybirds that eat aphids. Others are harmful, like locusts and aphids that feed on plants.

Birds

Some birds can be helpful to plants, for example by dispersing their seeds, and eating plant pests such as aphids. Others harm plants by feeding on them.

Questions

1 (a) What is meant by 'the environment'?
    (b) Write brief notes on 'Environmental factors that affect plants'.

2 Name two 'soil factors' and explain how they affect the growth of plants.

3 Name two 'climatic factors' and explain how they affect the growth of plants.

4 Name two 'biotic factors' and explain how they affect the growth of plants.

16

INTRODUCTION TO SOIL

When you have read this chapter you will be able to:

- *write a list of the things that make up soil (**constituents** of soil).*
- *explain where the constituents of soil come from.*
- *explain the weathering of rocks.*
- *describe the effects of three weathering agents.*
- *list the three main types of soil, and describe their properties.*
- *explain what is meant by soil texture, and soil structure.*
- *say what is meant by the soil profile.*
- *explain the importance of the profile in agriculture.*

Soil makes your shoes dirty, and your mother sweeps it out of the house. It is just dirt, isn't it? It cannot have much value. Can you think of any reason why we should regard soil as valuable or important?

You know, of course, that soil is actually very useful to us. The roots of plants grow in it, and get food from it. Plants could not possibly live without soil to grow in. Animals too depend on the soil, because they eat the plants that grow in it. And we ourselves could not survive if there were no soil, for the same reason.

You do not have to think long to realize that soil is the farmer's basic resource. Without it there could be no agriculture, no plants, no animals and no people! So really, there are not many things *more* important than soil.

How soil is formed

You can tell a lot about where soil comes from just by looking at it closely to see what is in it.

Fig. 16.1 Examining soil.

PRACTICAL WORK

Experiment To examine a sample of soil

1   Take some soil from the garden and examine it closely using a lens if you have one available.

2   Are the particles all the same size?

3   Notice their different shapes and colours.

4   See if there are any plant roots or small animals in it, or any pieces of dead animals or plants.

Soil particles can be separated into their different sizes by putting them in water, shaking them, and then letting them settle to the bottom of the water.

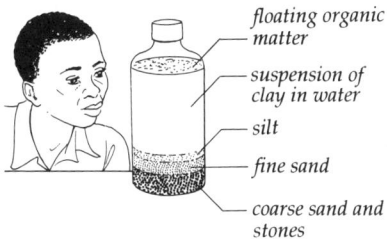

floating organic matter
suspension of clay in water
silt
fine sand
coarse sand and stones

Fig. 16.2 Separating the particles of soil.

PRACTICAL WORK

Experiment To separate soil particles in water, and examine them with a lens

1   Put three handfuls of dry soil into a big plastic bottle.

2   Pour water into the bottle until it is nearly full. Watch to see if any bubbles of air come out of the soil.

3   Cover the top of the bottle with your hand and shake it up, to mix the soil and water.

4   Stop shaking. Put the bottle on the table and watch carefully as the soil sinks to the bottom.

5   Which particles sink first? (Big ones or small ones.)

6   Do all the particles sink?

7   Does the water look muddy?

8   Why?

9   Do some things float on the surface of the water?

10   Describe the things that float on the surface of the water. Can you see pieces of dead plants, old roots or dead insects? These things make up the **organic material** in the soil which decomposes (rots) to form **humus**. Humus makes soil look dark brown.

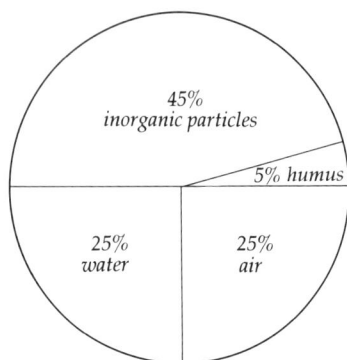

Fig. 16.3 *Things that make up the soil: pie chart showing soil constituents by volume.*

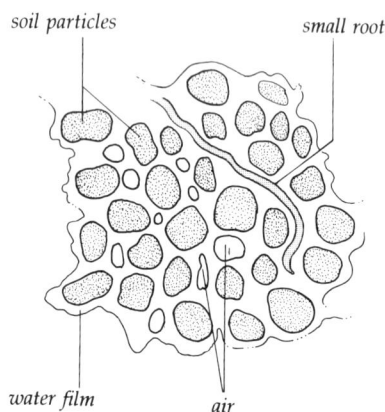

Fig. 16.4 *Soil particles highly magnified.*

11 *Leave the bottle overnight until the water is clear. Look at the soil particles that have now settled on top of the first layers. Are they large or small?*

12 *Draw a diagram of the experiment and label it to show the layers of soil particles.*

Things that make up soil

This pie chart shows the constituents of soil. This means the things that are present in the soil.

1 *Inorganic particles.* These come from rocks that have been broken down, or weathered. They make up about 45% of the volume of soil, (45 parts out of every 100). Plant roots grow in the spaces between these particles.

2 *Air* is also present in the large spaces between the particles. It takes up about 25% of the total volume. Air in the soil allows the organisms that live in the soil to breathe, and also helps plants to take up nutrients. Nutrient uptake by root cells is an active process requiring energy which comes from respiration.

3 *Water* is present in the soil as a 'film' clinging to the particles. Water is also held by humus, (the rotting remains of plants and animals). Water can form about 25% of the volume of soil.

4 *Plant nutrients.* The inorganic particles of the soil break down very slowly. As they do this, they produce **inorganic salts** that plants use as food. These salts dissolve in water, and plant roots take them in as they take in water.

5 *Humus.* This is the organic material in the soil, made from decomposed (rotted) remains of dead plants and animals, and from the waste products of living animals. It forms about 5% of the total volume. Humus releases inorganic salts, providing plant nutrients. It is also very good at:

- absorbing water

- holding water

- binding sand particles together

- separating clay particles
- providing food for soil organisms.

6 *Living organisms.* Many creatures live in the soil, such as earthworms, beetles, ants and termites. There are also many fungi that live in the soil, and enormous numbers of bacteria. These are called **micro-organisms**. A single gram of soil may contain about five million bacteria. The fungi and bacteria feed on the remains of dead plants and animals, and cause them to decompose.

Where soil comes from

Soil comes from rocks that have been broken down into very small fragments. But it is not easy to see *how* big, strong rocks can be broken down to make the tiny particles we see in soil. This happens as a result of weathering of rocks.

Weathering of rocks

Rocks are being broken down all the time by the action of water, wind and other factors. This breaking down process is called the **weathering** of rocks, and the things that break down rocks are described as **weathering agents**. New soil is being formed every day, but very, very slowly. The soil we see today has taken many thousands of years to form.

There are three groups of weathering agents:

- **Physical** weathering agents – like water, wind and temperature.

- **Biological** weathering agents – like plants and animals.

- **Chemical** weathering agents – these break up rocks by means of chemical reactions.

two pieces of sandstone being rubbed under water

Fig. 16.5 *Making soil particles.*

Fig. 16.6 *Comparing*

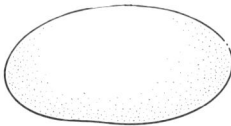

a) a smooth stone from a stream, with

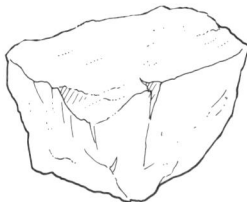

b) a sharply broken piece of stone

Fig. 16.7 *The Grand Canyon, formed by the Colorado River. What has happened to these rocks?*

Physical weathering agents

Water, wind and temperature changes can cause rocks to break down. This is how it happens.

Water as a weathering agent

1 *Running water*, as found in streams and rivers. Pieces of rock and stones are carried along by the water. The stones knock against each other and against the bed of the stream. Small pieces get broken off, and become soil particles.

PRACTICAL WORK

Experiment To break down rock into small particles

1 *Take two pieces of soft rock, such as sandstone. Rub them together under water in a basin or bucket. Be careful to keep your fingers out of the way.*

2 *Keep rubbing until you have produced some small pieces.*

3 *Examine these pieces with a lens, or under the microscope. Compare them with the soil particles you have already seen. Do they look the same?*

4 *What can you conclude from this experiment?*

Experiment To compare stones from a stream bed with a freshly broken stone

1 *Examine a smooth stone taken from the bed of a stream, and compare it to a freshly broken stone with sharp corners.*

2 *What has happened to the smooth stone?*

3 *Where has the material gone that was worn away from it?*

---

2 *The sea.* Waves on the seashore cause pieces of rock and stone to rub together and wear each other away, just in the same way as stones in a stream. Pebbles on a beach are always rounded.

Wind as a weathering agent

Strong winds wear away rocks by blowing sand against them. This happens most dramatically in desert places. If you have ever felt the wind blow sand against your legs, you know how it stings. Sand can strike rocks with force, and slowly wear them away.

Temperature change as a weathering agent

Heat makes rocks expand (get bigger), and cold makes them contract (get smaller).

The forces of expansion and contraction are very great. Rocks can crack when they are heated by the sun and then cooled by rain.

In desert areas there are often great changes in temperatures from daytime to night-time, especially when there is no cloud cover.

In cold climates water turns to ice in winter. When ice forms, it expands, (gets bigger). The force of this expansion is very great. Water in cracks between the rocks freezes, expands, and causes the rocks to break.

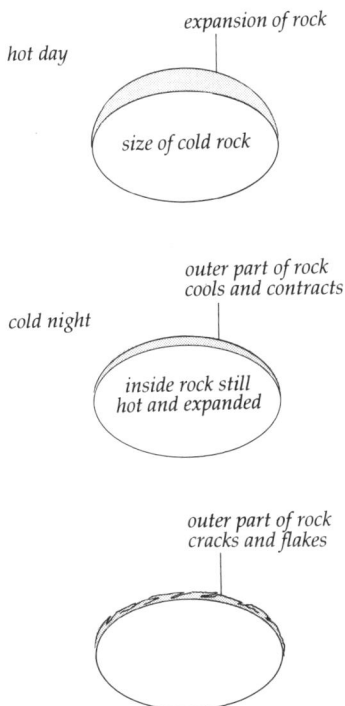

Fig. 16.8 Rocks in the Arizona Desert, worn away by wind and sand.

hot day

expansion of rock

size of cold rock

cold night

outer part of rock cools and contracts

inside rock still hot and expanded

outer part of rock cracks and flakes

Fig. 16.9 Heat and cold can cause cracking of rocks.

PRACTICAL WORK

Experiment To observe the expansion of freezing water

1 Take a glass bottle with a tightly fitting screw top. Fill it completely with water.

2 Put the full bottle into the freezing compartment of a refrigerator. Leave it overnight.

3 Examine the bottle the next day to see if it has been damaged by the formation of ice.

4 Explain any changes you observe.

Biological weathering agents

Plants as weathering agents

Tree roots grow into cracks between rocks and break them up.

Fig. 16.10 What will happen when the water freezes?

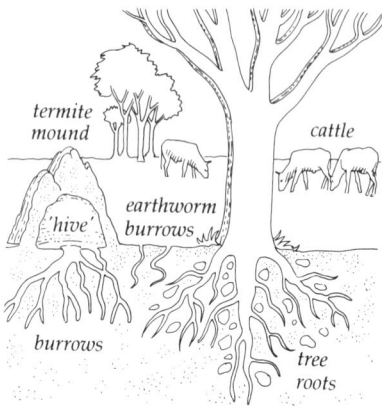

Fig. 16.11 Plants and animals cause weathering of rocks.

Animals as weathering agents

Earthworms, termites and other burrowing animals make tunnels under the ground. They help to break up soil into smaller particles, and also move subsoil particles to the surface. You can see this by examining worm casts on the soil surface. The soil particles have been ground together in the digestive system of the worm, and then passed out again. Note how fine the particles are.

Large animals, like cattle, help to break down soil particles when they walk over the soil. They cause the particles to rub together.

Chemical weathering agents

Rocks can react chemically with oxygen and carbon dioxide in the air:

- **Oxygen** can combine with inorganic particles in the rocks, such as iron, and this helps to break them down.
- **Carbon dioxide** in the air combines with rainwater to make a weak acid, (carbonic acid). This can slowly dissolve limestone and chalk rocks. If a piece of limestone is put into dilute hydrochloric acid, bubbles of carbon dioxide can be seen coming from it, showing that a chemical reaction is taking place. This is similar to the action of carbonic acid, present in rain, on limestone. A statue or a building made of limestone will slowly crumble away, due to this same type of chemical action.

Types of soil

There are three main types of soil – **sand**, **clay**, and **loam**. The following table shows the properties of each soil type. Notice how their properties depend, to a great extent, upon the size of their particles.

	SANDY SOIL	CLAY SOIL	LOAM SOIL
	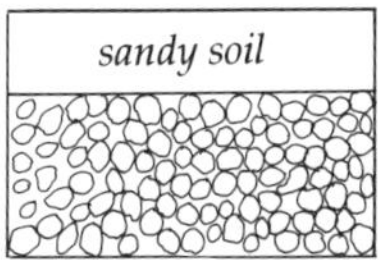 sandy soil	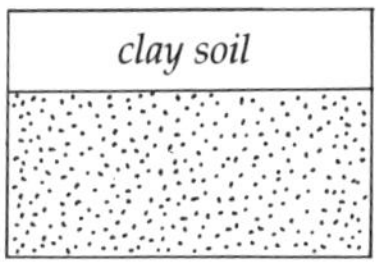 clay soil	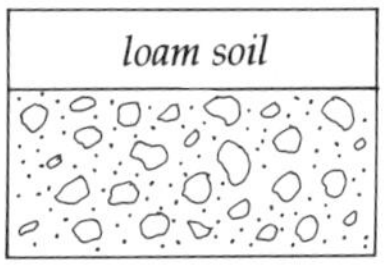 loam soil
Particle size	Large particles give soil a coarse texture.	Very small particles give soil a fine texture.	A mixture of large and small particles gives soil a moderately fine texture.
Water-holding capacity	Holds very little water. Plants may die very quickly in dry weather.	Holds a lot of water.	Holds enough water for plants.
Drainage	Drains quickly.	Drains very slowly.	Drainage is 'optimal', i.e. not too fast or too slow.
In dry and wet seasons	Plants suffer from a lack of water in dry weather.	Plants may be waterlogged in wet seasons.	Plants usually have an adequate supply of water in the dry and wet seasons.
Root penetration	Roots grow through sandy soil easily.	Roots do not grow through clay soil easily.	Roots grow through loam soil easily.
Cultivation	Easy to work.	Heavy to work. Sticky when wet. Hard when dry.	Fairly easy to work.
Air content	Well aerated.	Poorly aerated.	Optimal aeration.

Fig. 16.12 Types of soil.

Fig. 16.13

a) Sand

b) Clay

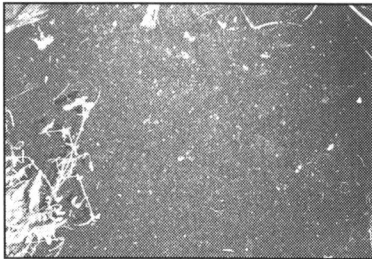

c) Loam

Which of these soils do you think is best for agriculture? Give reasons for your choice.

Soil texture

The term **soil texture** describes the proportions of sand and clay particles in a soil. In other words, the texture of a soil depends on the size of the particles in it. A soil with a lot of sand in it has a **coarse texture**. The particles are big. A soil with a lot of clay in it has a **fine texture**. The particles are very small.

Most soils are a mixture of sand and clay particles. They are loams, and their properties are shown in the right-hand column of Fig.16.12. Note that the *amounts* of sand and clay in these soils can vary. If they have a lot of sand, they are called **sandy loams**. If they have a lot of clay, they are called **clay loams.**

The best soils are **medium loams.** They drain easily, (because of the sand present in them), but they hold on to enough water, (because of the clay present in them).

PRACTICAL WORK

Experiment To compare the water-holding capacity and drainage rate of different soils

1 *Set up three measuring cylinders with filter funnels as shown in Fig.16.14.*

2 *Weigh 50 g portions of dry* **sand**, **clay** *and* **loam**, *and put them into the three funnels.*

3 *Pour 30 ml of water into each funnel in turn, and let it drain until it stops coming through.* How long *does it take for the first drop of water to drain through into the measuring cylinder?*

4 *After 10 minutes, measure* how much *water has drained through, and calculate how much has been* held *by the soil.*

5 Compare *the rate of drainage and the water-holding capacity of the three soils by copying and completing this table:*

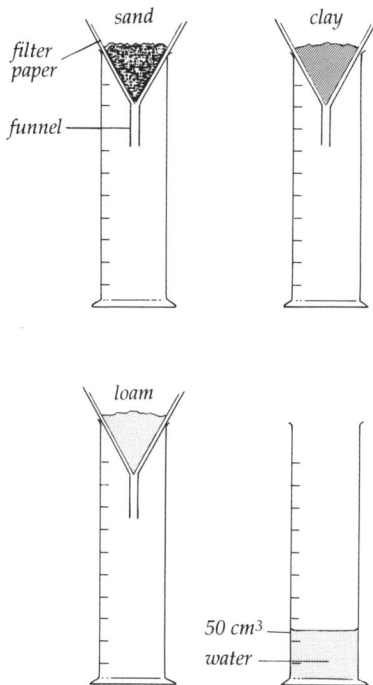

sand

clay

filter
paper

funnel

loam

50 cm³
water

Fig. 16.14 *Testing the water-holding capacity of soils.*

Sample	Amt. of soil sample (g)	Amt. of water poured into soil sample (ml)	Time for first drop of water (secs)	Amt. of water drained after 10 minutes (ml)	Amt. of water retained in soil (ml)	Type of soil (sand, clay, loam)
A						
B						
C						

6 Draw a diagram of the experiment.

7 Write down what you saw, and what you found out.

8 What would happen to plants growing in each of these soils if it did not rain for some time?

the soil sample
being rolled

Fig. 16.15 *The 'sausage' or 'soil worm' test.*

A quick 'field test' for soil texture

To find out how much sand and clay are present in a soil, do the 'sausage' test, sometimes called the 'soil worm' test.

1 Take a handful of moist soil, (wet, but not *too* wet).

2 Roll it between your hands, and try to make a 'sausage':

   – a **sandy soil** will *not* make a sausage at all

   – a **loam soil** will *just* make a sausage, but it begins to break up

   – a **clay loam** *easily* makes a sausage

   – a **pure clay** makes a sausage, and you can *wrap* it round your finger without breaking it.

3 Soils of different textures *feel* different when you rub them between your finger and thumb:

   – a **sandy soil** feels gritty

   – a **clay soil** feels smooth

   – a **loam** soil feels crumbly.

4 What can you say about the texture of your soil?

Soil structure

The **structure** of a soil depends on how the soil particles cling together due to the presence of humus. This has an effect on how the soil behaves. There are basically just two types of structure:

1 **Single grain structure** where the soil particles are separate from each other. For example, in river sand the particles do not stick together, even when they are closely packed.

Soils with a single grain structure have the following properties:

- poor water-holding capacity
- free drainage
- leaching of plant nutrients (plant nutrients are carried away in the drainage water)
- low numbers of soil organisms (like bacteria and fungi).

2 **Crumb structure** where the soil particles stick together in **crumbs** or **aggregates**. They are held together by humus, and look like brown 'breadcrumbs'. For example, a good garden soil has a crumb structure and it has the following properties:

- good aeration (air can get into the soil)
- good drainage (water can pass through the soil)
- good water-holding capacity (the soil holds on to a lot of water)
- plenty of healthy soil organisms
- it is easy to dig (it is not heavy or sticky).

The soil profile

If you have ever looked at a cutting on a hillside, or at a pit being dug, you will notice that the soil does not look the same – lower layers look different from the surface. What you are seeing is the profile of the soil, a section cut vertically down through it.

If you observe carefully, you will see that there appear to be several layers of different coloured soils. These layers are called **horizons**. The soil near the surface is usually

topsoil

subsoil

parent material

Fig. 16.16 *Examining a soil profile.*

darker in colour than the soil below, due to the presence of humus (decomposed remains of plant and animal matter). This horizon also contains the most plant roots.

The soil at the surface is know as **topsoil**, and in the tropics it varies in depth from 20–30 cm. The topsoil is the most important part of the soil for farmers. It is well aerated, rich in nutrients and has all the other conditions necessary for good plant growth. So it is described as fertile.

The horizon immediately below the topsoil is the **subsoil**. This is lighter in colour than the topsoil and contains more rocks. It is poorly aerated because it is more compact. Because the subsoil usually contains less readily available nutrients than the topsoil, and lacks the other good characteristics mentioned earlier, it does not provide good conditions for plant growth. It is infertile.

Below the subsoil you may see large rocks. These are known as **parent rocks**. These rocks were broken down by weathering and produced the inorganic soil that gave the soil its physical and chemical characteristics. This is why they are called 'parent' rocks.

You may also observe several sub-horizons. However, for most purposes, the layers may be grouped into one of the three main horizons:

- topsoil
- subsoil
- parent rocks.

PRACTICAL WORK

1 *Collect samples from each layer of a soil profile.*

2 *In a two-litre plastic bottle, construct a model of a soil profile using the samples you have collected.*

3 *Describe what you think will be the result for a farmer if the topsoil is removed, leaving the subsoil exposed (for example, by the action of man during land-clearing operations or by the action of wind or water).*

4 *How do you think the farmer can change the condition of the subsoil?*

1 The topsoil

- often dark in colour, due to **humus**
- depth from 5 to 30 cm
- contains plant roots
- is usually fertile
- gives plants water, nutrients and support
- should have a good crumb structure
- should be well drained and aerated

2 The subsoil

- lies below the topsoil
- lighter in colour, not much humus
- has never been cultivated
- not many plant roots or soil organisms present
- often not well drained or aerated
- deep rooted plants can get water and some nutrients from subsoil

3 The parent material

- has not been influenced by soil-forming processes

4 The parent rock

- lies deeper still
- sometimes called the bedrock
- can be very deep under the ground
- the rock from which the soil was formed
- boreholes sometimes reach this rock
- may contain water

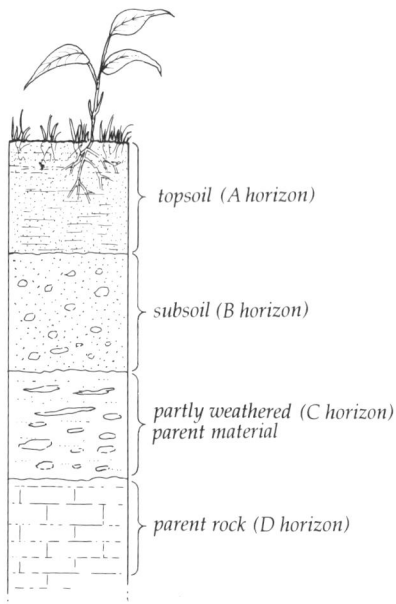

Fig. 16.17 The horizons within a soil profile.

Fig. 16.18 Erosion of the topsoil.

The importance of the topsoil

- The topsoil is extremely important to farmers. It is their basic resource since the crops or vegetables cannot grow without it.
- The fertility of topsoil can be increased by adding manures and fertilizers.
- Topsoil can be eroded. It is exposed to rain and wind.
- If topsoil is lost, no farming can be done in that place again.
- This is why we must look after and conserve our topsoil.

The Eleventh Commandment

You will inherit the holy earth as a faithful landlord, conserving its resources and its productivity from generation to generation.

Guarding and saving its fields against erosion, its vivid waters from drying, and its forests against desolation, so that your decendants will enjoy the abundance forever.

If anyone fails as landlord of the land his fields will turn to sterile and rocky dirt, and to ruining cracks, and your decendants will decrease and will live in misery or will be erased from the face of the earth.

Author: W. Loudermilk
Translator: Salazar

Questions

1   (a) List six constituents of soil.

   (b) Where does each constituent come from, and why is it important?

   (c) Draw a circular pie chart to show the approximate percentages of the main constituents by volume.

2   (a) Name the three main groups of weathering agents that cause rocks to break down.

   (b) Describe, with the help of a diagram, an experiment to separate soil particles in water.

   (c) What conclusions can you draw from this experiment?

3   (a) Draw a labelled diagram of soil particles highly magnified, with air and water present between them.

   (b) Why is it important for air to be present in soil?

4   Describe four effects of humus in the soil and explain why each one is helpful to plants.

5   (a) What is meant by 'soil texture'?

   (b) Describe, with the help of a diagram, an experiment to compare the water-holding capacity of sand and clay.

17

CLIMATIC FACTORS

When you have read this chapter you will be able to:

- *explain the difference between weather and climate.*
- *list climatic factors measured at a weather station.*
- *for each factor:*

    - describe the instruments used

    - demonstrate how the measurements are carried out

    - explain how the factor can affect plant growth

    - say how farming methods are adapted to cope with it.

- *explain the advantages of weather forecasting.*
- *draw a climate graph for your area.*

Weather and **climate** are not the same thing. When we speak about the **weather**, we mean the changes taking place from day to day in our environment – sometimes it rains, sometimes the sun shines, sometimes the wind blows, and we call all these 'changes in the weather'.

The **climate** includes *all* the changes of weather taking place throughout the year. For example, in the Caribbean, as you know, there is a wet season and a dry season during the year. The climate of an area can really only be judged from averages of records kept over a period of not less than thirty years.

Each part of the world has its own climate, depending on where it is on the earth's surface. Places near the equator are warmer than those further away. Places near the sea usually have more rain than those in the middle of continents. Places at high altitudes have lower temperatures than those at sea level.

We cannot change the climate in the place where we live, but there are two things we *can* do:

Fig. 17.1 Rain coming!

1   *We can study the climate* and learn how best to live with it, i.e. we can:

- Take measurements and keep records of climatic factors as they change throughout the year – factors such as temperature and rainfall, wind speeds and the humidity of the air.
- Study the effects of climatic factors on plants, to see which crops will grow best in our area.

2   *We can adapt our farming methods* to take advantage of the climate we have, i.e. we can:

- Follow farming practices that will fit in with the climate.
- Learn which seasons of the year have temperatures and rainfall which are most favourable for growing crops.
- Learn which crops grow best in our own climate.
- Grow these crops at the times when the climate suits them best.

Measuring the climate

In the next few pages you will see how to measure climatic factors, and how you can make instruments of your own for measuring them. The effects of each climatic factor on agriculture are also discussed.

Fig. 17.2 A weather station.

PRACTICAL WORK

Measuring climatic factors – making a weather station

1   *Try to set up a weather station to measure and record the daily variations in climatic factors.*

2   *You will need:*

- *A small, open area of flat land, say 5 m square, away from trees and buildings.*
- *Instruments for measuring the weather (these are described in this chapter).*
- *A fence around your weather station, to keep animals out.*

3 *The instruments used in a real weather station are expensive, but you can quite easily take useful measurements with simple home-made equipment. Your teacher can help you to set up your instruments and show you how to keep records.*

4 *Take readings* at the same time *each morning before the sun becomes hot. Try to measure the following weather factors:*

- *rainfall*
- *wind speed*
- *hours of sunlight*
- *humidity of the air*
- *air temperatures in the shade*
- *evaporation rate.*

5 *Share the work of keeping the records among the groups in the class and take turns. Let one group of students take readings for a week at a time. When your turn comes, don't forget! You need to take readings every morning.*

6 *If there is a real weather station near to your school, you may be able to go and see it working.*

7 *As you keep your records, construct a* **climate graph** *for the year, like the one in Fig. 17.3. As you fill in your graph, it will show you how temperature and rainfall in your area change from month to month.*

8 *Official records have been kept for many years. By taking averages from these, it is possible to see the normal, average temperatures and rainfall for each month of the year. Long-term records like these can tell you what weather to expect, and how frequently exceptional events, such as storms and hurricanes, are likely to occur.*

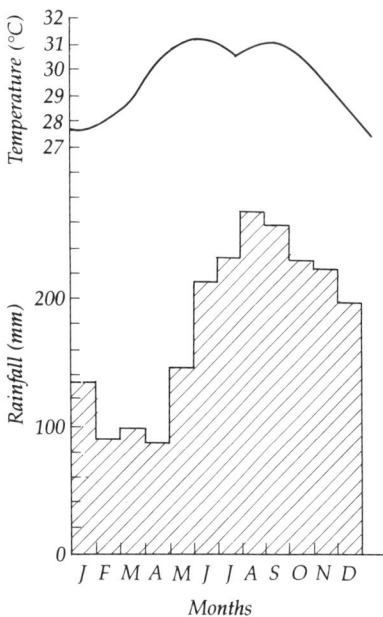

Fig. 17.3 *A climate graph for Soufrière, St Lucia.*

Measuring rainfall

A **rain gauge** is used for this.

The rain gauge is basically just a funnel and a jar for collecting rain drops. The collecting jar is usually placed below soil level to keep it cool and prevent evaporation, but you can easily lift it out. The amount of rain falling

Fig. 17.4

a) A rain gauge

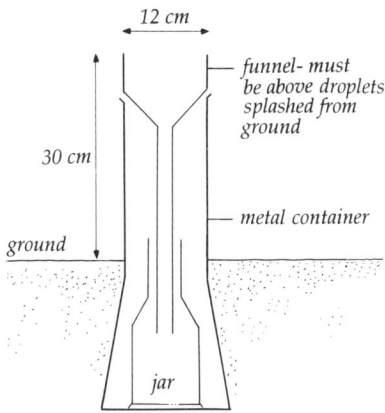

b) A section through a rain gauge

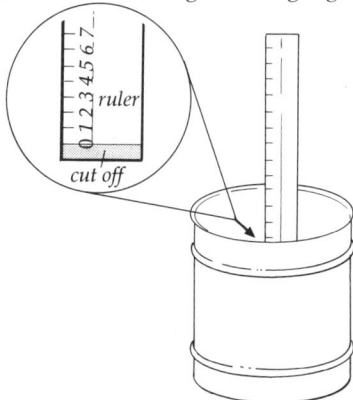

Fig. 17.5 A home-made rain gauge.

in one day is found by pouring the water collected into a measuring cylinder. This is done every morning, when readings are taken. Then the gauge must be emptied ready for the next day's collection. Rainfall is expressed as the *depth* of water in millimetres that falls on the ground.

The **total rainfall** for the year is an important feature of the climate, because it controls what crops can be grown in the area. The **rainfall distribution**, (i.e. the times of the year when rain falls), must also be known, so that planting can be done at the best times.

PRACTICAL WORK
Making a rain gauge

You can make a simple rain gauge from an old tin and a ruler, as shown in Fig.17.5.

1 *Cut the ruler with a saw exactly on the zero mark.*

2 *Leave the tin outside in the open, and measure the rainfall each morning by putting the ruler into the water to see how deep it is. This gauge will not be quite as accurate as a proper rain gauge, but it is good enough to give useful results.*

3 *Note that 1 mm of rainfall is equivalent to 1 litre of water falling over 1 square metre of land. If you know the rainfall in mm, you can work out the volume of water that the land has received. For example:*

- *1 mm of rainfall is equivalent to 1 litre falling on each square metre of land*
- *10 mm is equivalent to 10 litres per square metre*
- *20 mm is equivalent to 20 litres per square metre of land.*

One hectare is 10,000 square metres. If just one mm of rain falls on a hectare, the total volume of water will be 10,000 litres!

The effect of rainfall on plant growth

The questions farmers ask are:

- **'How much rainfall will we have in the year?'**

There must be a *sufficient total amount* of rain for crops to grow.

- **'How is the rainfall distributed through the year?'** It is not much help if all the rain falls within a few days, and then there is a drought. Plants cannot make use of too much water all at once, and they cannot survive for long without any rain. Most crops need an average of 7 mm of rainfall per week for normal growth. The rainfall must enable crops to complete their life cycle, from germination right through to harvest. Crops do best when there are regular falls of rain, well distributed (spread out), throughout the growing season.

- **'Will there be storms or hurricanes?'** Very heavy rains and wind can do a lot of damage to soil and crops. Water cannot soak down into the ground fast enough, and begins to run over the surface causing erosion. Plants are beaten down. Flowers and fruits fall off trees.

- **'How reliable is the rainfall?'** Records from previous years give us a good idea of how much rain to expect and when to expect it. This is because the records show not just how *much* rain there was, but *when*, and *how often* it fell.

You can see that when farmers know the answers to these questions, they are in a good position to judge how useful and effective the rainfall will be in helping their crops to grow.

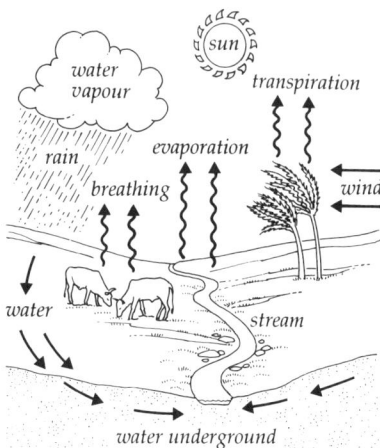

Fig. 17.6 *The water cycle.*

The water cycle

From the diagram in Fig. 17.6 you can see that water **evaporates** to become **water vapour** in the atmosphere. Water may evaporate from the sea, from wet ground, from plant leaves (in transpiration), from animals (as they breathe and perspire), from clothes drying on the washing line, or anywhere else where water is exposed to the air.

The air can hold an enormous amount of water, in the form of vapour. When it cools at high altitude, this vapour **condenses** to form clouds, and may fall to earth again as drops of rain. We call this movement of water in the environment the water cycle.

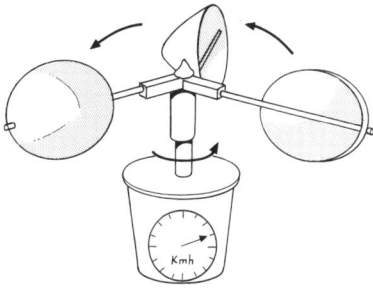

Fig. 17.7 An anemometer.

a)

b)

washing up
bottle

cut

c)

vanes tied on
to the spokes

Fig. 17.8 Making an anemometer.

Measuring wind speed

Wind is the movement of air over the earth's surface. It moves from a region of high pressure to one of low pressure, and that is when we feel a wind blowing.

The instrument for measuring the speed of the wind is an **anemometer**. It has three metal cups, mounted on a spindle that spins round when the wind blows. The faster the wind, the faster it turns. It works like a speedometer in a car, and has a dial showing wind speed in kilometres or miles per hour. The only trouble is, a proper anemometer is expensive.

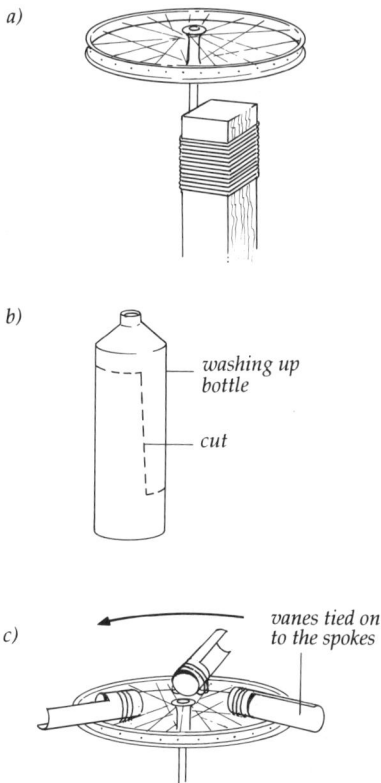

PRACTICAL WORK
Making an anemometer

This is not so difficult to do as it might seem at first.

1 *Find an old pram wheel and mount it vertically, still on its axle, on top of a fence post, as shown in Fig. 17.8.*

2 *Cut up some plastic bottles, (e.g. washing-up liquid bottles), to make the vanes of the instrument, and mount them vertically on the wheel. You could tie them on with wire or string.*

3 *Oil the wheel well, and check that it turns in the wind. You can make a mark on the wheel, with paint or a sticky label, so that the revolutions can be counted. Then you can compare the number of revolutions at different wind speeds.*

4 *To find the speed in kilometres per hour is a bit more difficult. You might like to try **calibrating** your anemometer by timing the movement of a balloon over a measured distance at different wind speeds. One group releases the balloon at a given signal and times how long it takes to travel, say, 10 metres. Another group counts the revolutions of the anemometer at the same time.*

If you know how long it takes for the balloon to travel 10 metres, you can multiply this by 100 to find its speed in km/hr. Then you can relate the revolutions per minute, (or R.P.M.), shown by your anemometer to actual wind speeds. Your teacher will help you.

The effects of wind on plants

Wind can have harmful and good effects. Here are some of them:

Bad effects	Good effects
1 **Blowing plants down**, especially when the wind is accompanied by heavy rain.	1 **Wind pollination** of cereals and grasses.
2 **Wind erosion** occurs if the soil is dry and and dusty, and there is no plant cover.	2 **Removal of excess moisture.** Wet plants are more likely to be attacked by fungi.
3 **Excessive evaporation** in hot, dry weather	When air moves over a plant the humid air surrounding it is carried away. This helps to keep the plant safe from fungal attack. Pruning of tree branches helps the wind to circulate and remove this moisture.
4 **Transpiration rates** in plants increase with the wind speed. If plants cannot replace the water they lose fast enough, they wilt, and may die.	

Farmers can control the effects of wind to some extent, by planting hedges or trees as **windbreaks,** and by spacing crop plants correctly so that the wind removes humid air from around plant leaves.

Measuring hours of sunlight

If you have ever tried burning a hole in a piece of paper by focusing the rays of the sun with a lens, you will understand how a **sunshine recorder** works.

Fig. 17.9 shows a sunshine recorder in operation. It records the times when the sun shines during the day.

At the centre of the recorder is a sphere of glass, and behind this is a curved piece of card. The glass acts as a lens. When the sun shines, it focuses a spot of light on the card, causing a burn mark. As the sun moves across the sky, the spot of light traces a burn mark across the card. If the sun goes behind a cloud, the burning stops. If it comes out, the burning starts again. Special cards are printed with the hours of the day marked on them. At the end of each day the card shows exactly when the sun shone, and for how long. A new card is put in place every morning.

Fig. 17.9 *A Campbell-Stokes sunshine recorder.*

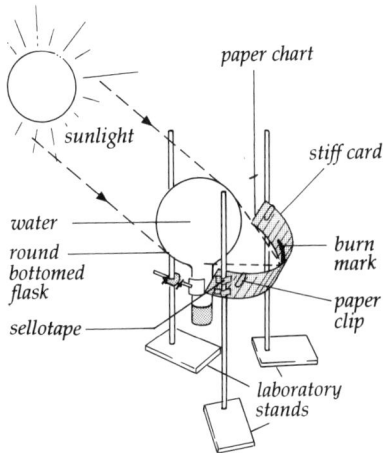

Fig. 17.10 A home-made sunshine recorder.

PRACTICAL WORK
Making a sunshine recorder

1 You can make a sunshine recorder from a spherical flask filled with water, held in place by a laboratory stand, as shown in Fig. 17.10.

2 Set up the apparatus and, by trial and error, make a spot of light fall on a card to cause a burn mark. You have now found the **focal length** of the spherical lens.

3 Now make a rectangular card of a suitable size, and set it up in a curved position so that it is **concentric** with the flask at just the right distance to receive a burn mark for one whole day. Concentric means having a common centre. It takes some experimenting to get this right, but when you have managed it, you have made a sunshine recorder.

4 As the sun moves round, mark the hours of the day on the card with a pencil, as though it were a sun dial. You need do this only once. Remove the card and draw in graduations neatly to show the hours and minutes.

5 Next, produce a lot of paper copies of the card using a photocopier, so that you can make a new record each day. Now reassemble the sunshine recorder, and each morning clip a new paper chart on to the inside of the card, using paper clips. Write the date on the chart.

The effects of sunlight on plants

Light is essential to plants for three main reasons. You may remember them from earlier lessons:

- **The formation of chlorophyll** – plants cannot form the green pigment **chlorophyll** unless they are exposed to the light.

- **Control of the direction of growth** – plants shoots grow towards the light. They compete with each other to reach the light. This is why it is important to space plants correctly, to give them all enough light.

- **Photosynthesis** – this is the process by which plants make sugar. It can take place only in the light.

anemometer

Stevenson screen
with thermometers

rain gauge

maximum thermometer

maximum temperature during last recorded period

small index pushed up by expanding mercury - stays in position when mercury contracts

mercury

minimum thermometer

small index pulled down by contracting alcohol - does not return when alcohol expands

minimum temperature during last recorded period

alcohol

Wet

°C °F
43 110
38 100
32 90
27 80
21 70
16 60
10 50
4 40
-1 30
-7 20
-12 10
-18 0

Dry

°C °F
43 110
38 100
32 90
27 80
21 70
16 60
10 50
4 40
-1 30
-7 20
-12 10
-18 0

muslin

container

water

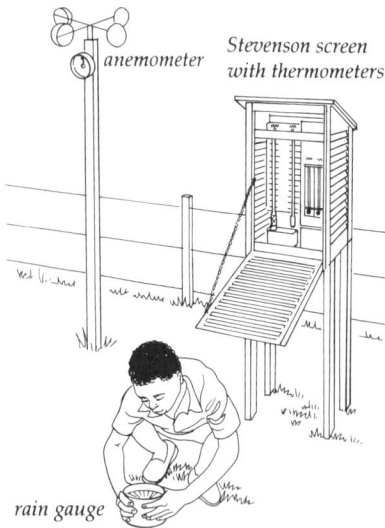

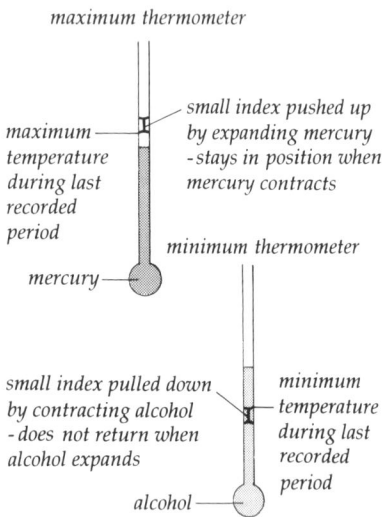

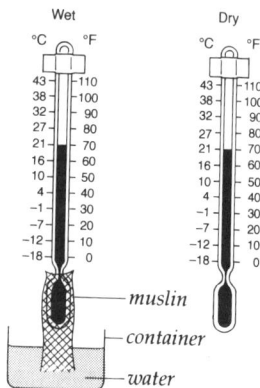

wet and dry bulb thermometer

Fig. 17.11 A Stevenson screen.

Light also affects the reproductive cycle of many plants, a process known as **photoperiodism**. This is related to the day length, i.e. the hours of sunlight and darkness. Pigeon peas and sorrel are two local crops which must have shorter hours of daylight to trigger their flowering phase. Researchers, however, have identified certain varieties which are **day neutral**, i.e. they do not have to have shorter days to initiate flowering and fruiting. These are generally referred to as **dwarf varieties** because they begin to produce flowers and fruit within three months when there is not yet a lot of vegetative height growth, e.g. Chag Pearl, U.W. 10, and dwarf sorrel.

Some crop plants (known as **sun-loving plants**) need as much sunlight as possible to produce effectively. Therefore all sources of shade should be removed from the areas where the plants are growing. Plants like sugar cane, tomato, pumpkin and cabbage need maximum sunlight. The plants should also be spaced correctly to prevent them from shading each other.

Some plants (known as **shade-loving plants**) do not require as much light. If they receive too much light, their growth and production falls. They must either have natural shade, provided by trees, or be self-shaded (i.e. shading each other), or artificially shaded. Tree crops, such as cocoa and coffee, need a certain degree of shading to achieve maximum productivity. Many of the exotic tropical floriculture and herbage plants require shading to prevent discolouration of petals, or chlorosis (yellowing) of the leaves due to sunburning.

Farmers cannot alter the times when the sun shines, but they can try to give their crops the right exposure to light by spacing plant rows correctly, so that the plants will not shade each other too much. They can also provide shade where it is needed.

Measuring temperature

A real weather station has a special louvred box, called a **Stevenson screen**, for shading the thermometers (Fig. 7.11). This allows the thermometers to be exposed to the air, but not to the sun. Air passes freely through the screen, which is painted white to reflect the sun's rays. The screen has a door that can be opened easily, and is placed at a convenient height for reading the thermometers.

You may be able to make a Stevenson screen from some louvred window shutters, but if not, you can still take useful readings. One thing you must do is keep your thermometers in the shade.

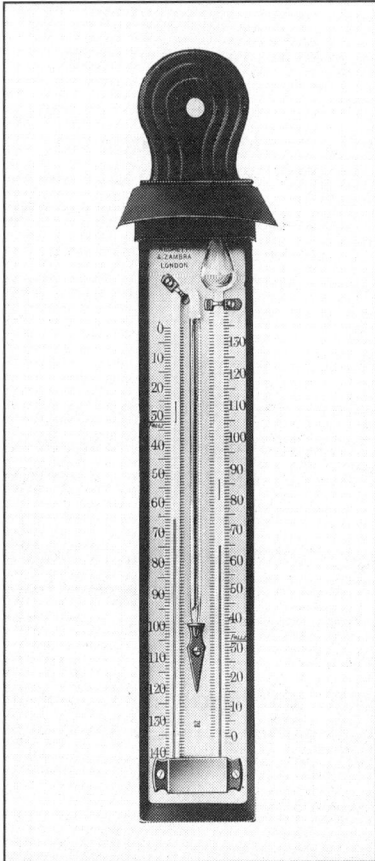

Fig. 17.12 A maximum and minimum thermometer.

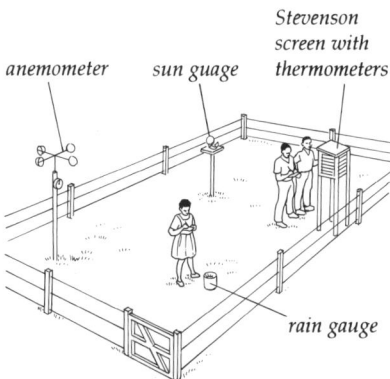

anemometer   sun guage   Stevenson screen with thermometers

rain gauge

Fig. 17.14 Recording the weather.

PRACTICAL WORK
Reading the thermometers

1 *The best instrument for measuring air temperatures is the* **maximum and minimum thermometer**. *Your teacher will show you how to use it. This thermometer shows:*

- *the temperature at the present time*
- *the maximum (highest) temperature reached during the previous day*
- *the minimum (lowest) temperature reached during the previous night.*

2 *Read the thermometer each morning, and keep records by plotting a graph as shown by Fig.17.13. The difference between the highest and lowest temperatures is called the* **daily range**, *or the* **diurnal range**.

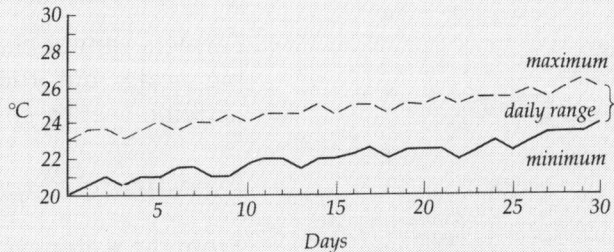

Fig. 17.13 A graph of temperatures showing the daily range.

3 *Remember to:*

- *Keep thermometers in the shade so that they record the true temperature of the air. What would happen if you placed them in the sun?*
- *Read the thermometer at the same time every morning. Why?*
- *Join maximum and minimum points on the graph to produce curves showing how temperatures vary over time.*

• *Keep your record going for a year at least, to observe the seasonal variations in temperature.*

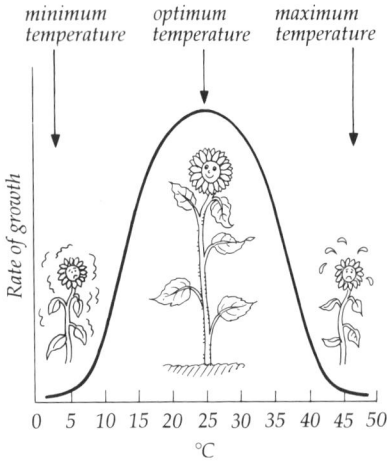

Fig. 17.15 How the rate of plant growth varies with the temperature.

The effects of temperature changes on plants

Plants respond to temperature changes by growing faster or slower. This is because the **metabolic rate** of plants, (the rate at which chemical reactions take place inside them), depends on the temperature. Generally speaking, plants like it to be warm, but not *too* warm. The graph in Fig. 17.15 shows the relationship between temperature and the rate of plant growth.

Note what the terms on the graph mean:

* The **minimum temperature** for growth is 5 degrees Celsius. Below this temperature there is no growth at all. Fortunately, temperatures do not fall as low as this in the Caribbean.

* The **optimum temperature** for growth is between 20 and 25 degrees Celsius. Most plants grow best within this range. As the temperature rises higher, growth slows down again.

* The **maximum temperature** for growth is about 40 degrees Celsius. If it gets as hot as this most plants begin to die. Only cacti and other desert plants are adapted to withstand extreme heat.

Farmers cannot change the temperature of the air on a big scale, but they can protect young plants from extreme heat by shading, mulching and irrigation.

Measuring evaporation and humidity

From the **water cycle** (see Fig. 17.6), you saw how water evaporates to form water vapour in the air. Water **changes state**, from liquid to gas. The rate of evaporation increases greatly when the sun shines brightly, or the wind blows strongly.

In areas where very rapid evaporation causes problems, weather stations may have an **evaporation pan**. This measures the exact rate of evaporation accurately. Farmers can then match the amount of irrigation water they apply to the soil to the amount of water lost by evaporation.

Fig. 17.16 An evaporation pan.

Humidity

Water vapour makes the air **humid**, or moist. Here are some facts you should know about the humidity of air:

- When the air is holding as much water vapour as it can, it is said to be **saturated**.
- Warm air can hold more water vapour than cold air before becoming saturated.
- If warm air, near to the saturation point, cools down, then **condensation** takes place and it rains.
- The amount of moisture in the air is expressed as the **relative humidity** of the air. This is defined as the amount of moisture the air is holding, compared with the amount it *could* hold at any given temperature if it were *completely saturated*. If you read that again slowly you will see what it means.

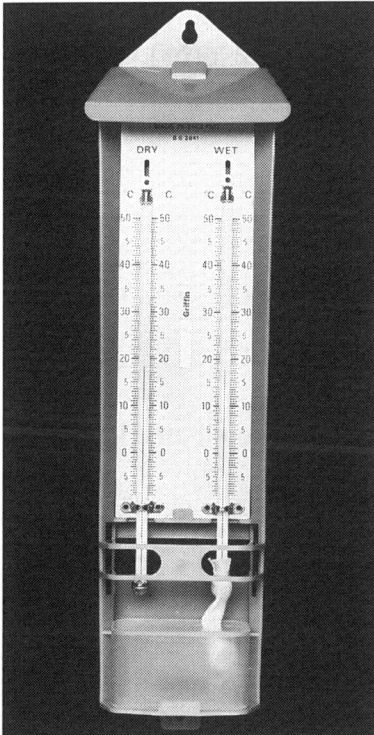

Fig. 17.17 A wet and dry bulb hygrometer for measuring the relative humidity of the air.

PRACTICAL WORK
Measuring humidity

1 *The instrument for measuring the **relative humidity** of the air is called a **wet and dry bulb hygrometer**.*

2 *There are actually two thermometers making up this instrument, and they are kept together in a Stevenson screen, or in the shade. The dry bulb thermometer is a normal thermometer, giving the temperature of the air. The wet bulb thermometer has its bulb covered by a piece of muslin cloth. This cloth is kept damp by a jar of water with a wick coming up from it, as shown in Fig. 17.17. It is not difficult to make your own wet and dry bulb thermometers.*

3 *The thermometers work as follows:*

- *On a humid day when the air is saturated, the two thermometers give the same reading, but on a dry day the wet bulb thermometer will give a* lower *reading than the dry one. This is because it is cooled as water evaporates from the cloth.*
- *The wet bulb thermometer usually gives a lower reading than the dry bulb one.*
- *As the humidity of the air varies between completely dry and completely saturated, the difference between the readings also varies.*

> • *We can find the number of degrees difference between the thermometer readings by subtraction. This number can then be compared to special tables supplied with the thermometers, to find the* percentage relative humidity *of the air.*

Effects of humidity on plants

The relative humidity has a big influence on plants because of the way they are continuously gaining and losing water. For example:

A *high* relative humidity has harmful effects and good effects:

Bad effects	*Good effects*
1 **Encourages fungal diseases.** Most plant diseases are caused by parasitic fungi, that grow best in warm, humid conditions. Examples are potato blight, tomato blight, rust of cereals, fungal spot in beans. The risk of fungal attack is greatest when the weather is warm and wet. If plants are overcrowded by not being properly spaced in rows, a **micro-climate** can develop among the leaves, where the air is warm and moist. This can lead to fungal attack.	1 **Slows down evaporation.** This is good for young roots, as it keeps the soil moist.
	2 **Slows down transpiration.** Plant leaves and stems stay turgid (full of water). This gives them support and helps growth.

A *low* relative humidity also has harmful effects and good effects:

Bad effects	*Good effects*
1 **Increases the rate of transpiration.** Plants lose water faster than they can take it up from the soil. Then they **wilt**. Their stems and leaves droop, the rate of photosynthesis decreases and plants cannot grow.	1 **Reduces risk of fungal attack.** Fungi do not like dry conditions.
2 **Dries out the surface soil.** This is dangerous to young seedlings with shallow roots. We can try to protect young plants in nursery beds by watering and shading them.	

Coping with humidity changes

As the humidity of the air varies, plants have to cope with the changes. Farmers cannot alter the humidity of the air on a big scale, but they can reduce some of the unfavourable effects.

- Use a mulch on seedbeds to keep the soil moist.
- Space plant rows wide enough apart to allow free movement of air, and avoid fungal attack in humid weather.
- Grow varieties of crops that are resistant to fungal diseases.
- Irrigate crops when the weather is dry.

Answering questions about your own climate

When you have been keeping records for a while, discuss these questions with your classmates and see if you can answer them:

How is your own climate related to:

- your position on the earth's surface, (latitude and longitude)?
- your nearness to the sea?
- the altitude?
- the prevailing winds?

Fig. 17.18 *Checking the forecast.*

PRACTICAL WORK
Checking the weather forecast

The weather forecast on television is well worth watching. It is prepared daily by experts who have excellent instruments, and can carry out detailed observations from all sorts of places – on the ground, from ships and aircraft, high altitude balloons, and even from earth satellites. The forecast also shows you how whole weather systems are moving across your region, and at what speed.

When you have been keeping records yourself, you will understand the forecast better than most people.

1 *Watch the weather forecast, or look it up in the newspaper. Check it against your own weather observations made at school.*

2 *Note how changes in temperature, humidity and wind speed coincide with movements of large weather systems.*

3 *When you know from the forecast that a sudden change, such as a storm, is coming your way, take extra careful note of your weather station readings at these times to observe the changes.*

4 *Notice how sometimes your readings are different from the forecasters' predictions. This can happen when you are just on the edge of a large, moving weather system. Sometimes the forecasters are right about what is coming, but the weather system might move slightly slower or faster than they were able to predict.*

5 *It is nice to know that you can do some of the things that weather forecasters do, and that your own readings point to the same conclusions as theirs.*

Nature

We have neither Summer nor Winter
Neither Autumn nor Spring.

We have instead the days
When gold shines on the lush green canefields –
Magnificently.

The days when the rain beats like bullets
    on the roofs
And there is no sound but the swish of water in the
    gullies
And trees struggling in the high Jamaica winds.

Also there are the days when the leaves fade from off
    the guango trees
And the reaped canefields lie bare and fallow in the
    sun.
But best of all there are the days when the mango and
    the logwood blossom.
When the bushes are full of the sound of bees and the
    scent of honey,
When the tall grass sways and shivers to the slightest
    breath of air,
When the buttercups have paved the earth with yellow
    stars
And beauty comes suddenly and the rains have gone.

H.D.Carberry

Questions

1 (a) What is the difference between 'weather' and 'climate'?

   (b) Write brief notes on the climate of your own country under the headings:
   – Temperature changes during the year
   – Rainfall
   – Normal wind speeds
   – Exceptional wind speeds.

2 (a) What would be a suitable site for a weather station?

   (b) Draw a plan of a station to show the layout and the instruments it contains.

   (c) Explain briefly what each instrument does.

   (d) How often should observations be made at the weather station?

3 (a) Draw the instrument used to measure the humidity of the air, and explain how it works.

   (b) How could high air humidity have a harmful effect on a growing crop?

4 (a) Draw a diagram of a rain gauge and explain how it works.

   (b) Why do farmers prefer rainfall to be well distributed throughout the growing season?

5 (a) Draw a maximum and minimum thermometer and explain how it is used.

   (b) What is the optimum range of temperature for the growth of most plants?

   (c) What could happen to plants if the temperature rose above the optimum range?

   (d) What could happen to plants if the temperature fell below the optimum range?

18

BIOTIC FACTORS

When you have read this chapter you will be able to:

- *explain what is meant by the biotic factors in the environment.*
- *list the biotic factors.*
- *describe the effects of these factors on plants.*
- *say how farmers can help crop plants to cope with biotic factors.*
- *describe methods of cultural control of pests and diseases.*

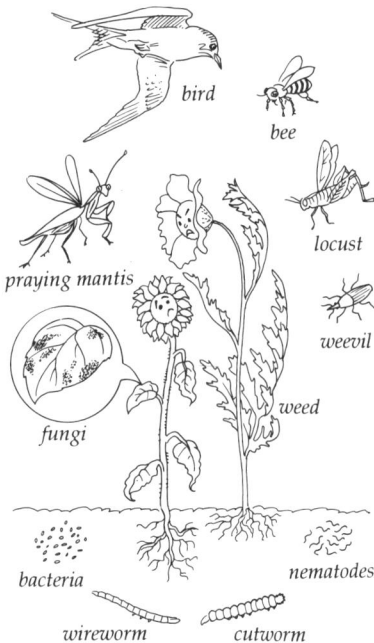

Fig. 18.1 *The biotic factors affecting plants.*

Plants share their environment with many other living organisms, and these can have important effects on plant growth. These other organisms are the **biotic factors** in the plant's environment.

The main biotic factors were mentioned briefly in Chapter 15, but now is the time to consider them in more detail. To remind you, these are the biotic factors:

- weeds
- insects (helpful and harmful ones)
- other animals and birds
- micro-organisms:
    - bacteria
    - fungi
    - nematodes
    - viruses.

We have to think about how each factor affects agriculture, and what farmers can do to cope with them.

Weeds

A weed is a plant growing in the wrong place. Never forget that since weeds are plants, they will compete with your vegetables and crops for:

- plant foods in the soil
- space in the plot
- water in the soil
- light.

Think of all the work you put into helping your vegetables and crops. You:

- prepare the land
- manure and fertilize it
- dig and rake it
- measure out plots
- carefully sow your seeds
- apply mulch
- water the plants
- and shade them.

You do all this *for your precious plants*, not for the weeds!

Weeds will try to take full advantage of all the good conditions you provide. They harbour pests and diseases that might spread to crops, *and* they make the garden look very untidy. Weeds are definitely the enemy!

Fig. 18.2 Composting weeds.

PRACTICAL WORK
Fighting weeds

Fortunately, weeds cannot fight a whole class of students! You can do a lot to protect your plants from them. You can:

1 *Cut weeds down with hoes between your rows of plants.*

2 *Pull them up by hand where they are growing between the plants in the rows. Be careful not to pull up the crop plants at the same time.*

3 *Put the weeds on the compost heap where they will decompose and make useful humus.*

4 *Never let weeds get bigger than your plants.*

5 *Never let them live long enough to produce flowers and seeds. If their seeds fall on the soil, there will be even more weeds next year.*

6 When you dig the garden, bury the small weeds. Turn them upside down into your trench. They will rot and make good compost.

7 Write in your diary the dates when you carried out weeding, and say why it was important to do it.

Collecting weeds

It will help you to learn about weeds if you make a pressed collection of them, as you did for wild plants, (see Chapter 4). This is how to do it:

1 Collect weeds from the garden and bring them back to the classroom. Try to find good specimens that show the flowers and fruits of each weed. Make a note of the date and place where you collected them.

2 Press them between sheets of newspaper, under a stack of heavy books. They will dry out in two to three weeks. Then you can mount them on white paper. Remember to change the paper daily to prevent rotting due to fungi.

3 When you mount the weeds, write the name of the plant on the page and the date and place where it was collected.

4 Display your collection of pressed weeds on the classroom wall.

Fig. 18.3 A display of pressed weeds.

Exercise

Copy and finish these sentences:

- Weeds steal _____, _____, _____ and _____ from vegetables.
- If weed seeds fall to the ground, there will be _____ weeds next year.

Insects

Insects are among the most successful groups of animals on the earth. They are small, but there are millions of them! 750,000 different species of insect have been described so far, and more are still being discovered. They live everywhere – in the soil, on plants, in water, on land, in the air, on trees, up mountains, under stones, on the backs of animals – wherever you go you are sure to find insects.

Fig. 18.4 *Bees help to pollinate an orchard.*

Fortunately, most of them get on quietly with their own lives and want nothing to do with us, but *some* insects do affect our crop plants quite a lot. These are the ones we need to think about now. There are helpful insects and harmful insects.

Helpful insects

Some insects, such as the ladybird and the praying mantis, are predators that kill and eat garden pests. If you see them, do not kill them.

Bees and many other insects are also very helpful because they pollinate flowers. As they feed on nectar, they move from flower to flower carrying pollen on the hairs of their bodies. Fruit farmers often keep hives of bees in the orchard because their trees will not form fruit unless their flowers are pollinated.

Harmful insects

Quite a lot of insects and their larvae do real damage to our crops. They have a bad effect on the plant environment. Insect pests are classified (grouped) according to the kind of damage they do. Here are some examples:

Type of pest	Examples	Damage caused	How to control
Biting pests	cutworm – caterpillars of the cutworm moth	bite off young plants at ground level	a poisonous cutworm bait
	grasshoppers and locusts	eat plant leaves	spray with contact or stomach insecticide
	bachac	eat plant leaves	place bait near their nest
Boring pests	maize weevil and bean weevil	larvae bore holes in stored seeds	fumigate food stores to kill larvae
Piercing and sucking pests	aphids, mealy bugs and rice stinkbug	pierce plant tissues and suck juices	spray with systemic insecticide

| Soil pests  | wireworm larvae of the click beetle | eat stems of young plants just below ground level | rotate crops |
| 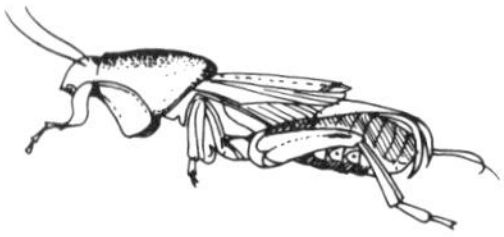 | mole cricket | feed on young roots | rotate crops, use poison bran baits, as for cutworms |

Fig. 18.5 *Harmful insects and how to control them.*

Often crop pests will attack just one type of crop. They are said to be **specific** to that crop. This means that crop rotation is often a very good way of controlling their numbers.

PRACTICAL WORK

Watching out for pests and diseases

1 *Look carefully at the leaves of your crop plants. Turn them over to see if any insects are living on them or eating them.*

2 *Larger pests, such as caterpillars and beetles, can be picked off by hand and crushed under your foot.*

3 *Look for any coloured spots on the leaves, caused by fungal disease.*

4 *Check the flowers and pods.*

5 *If you see any pests or diseases, tell your teacher straightaway.*

6 *Your teacher may decide to spray pests with a* **pesticide** *chemical that will kill insects, or a* **fungicide** *that will kill parasitic fungi.*

7 *Write down in your diary any pests or diseases you found. Draw a plant, showing the damage caused. Say what control measures were used.*

Exercise

Copy and finish these sentences:

- *Holes in leaves are caused by* _____.
- *Coloured spots on leaves are a sign of* _____.

Other animals and birds

Some animals and birds can be helpful to plants, for example by dispersing their seeds, and feeding on harmful insects. Others harm plants by feeding on them.

Larger organisms in the soil, such as earthworms, beetles and their larvae, are helpful. Their burrows let air and water into the soil, and improve drainage and aeration.

Micro-organisms

These are organisms of microscopic size, in the environment, which have effects on the growth of plants. Again, there are helpful and harmful micro-organisms.

Helpful micro-organisms

As you already know micro-organisms, such as **bacteria** and **fungi,** are present in the soil in large numbers. They feed on dead plants and animal waste, causing them to decompose. The products of decay are the nutrients needed by living plants. This means that soil organisms have a vital part to play in recycling plant nutrients.

Harmful micro-organisms

The micro-organisms that cause damage to plants belong to several groups. Here are some examples:

Micro-organism	Examples	Damage caused	How to control
bacteria	bacterial wilt	xylem vessels become blocked, plants wilt suddenly	no cure, grow other crops in rotation
fungi	most plant diseases are caused by parasitic fungi of different kinds	spots and moulds on leaves and stems	spray with a fungicide to stop the spread of disease, burn affected plants

nematode worms	potato root eel worm	millions of tiny thread-like worms destroy roots and tubers	fumigate soil, rotate crops
viruses	tobacco mosaic virus	yellow mottling of leaves	remove and burn affected plants, burn crop residues

Fig. 18.6 Micro-organisms and how to control them.

Whenever you are going to grow any vegetable or crop plant, it is always a good idea to check in advance on the specific pests and diseases that could attack it. Then you will be ready to take action to control any problems.

Cultural control of pests and diseases

All this information about biotic factors helps us to *know what to do* in the garden. Think of it from a plant's point of view. If you were a plant, how would you like it if you were exposed to attack from weeds, insects and diseases! You would be very glad if someone could protect you from them.

Here are some things you can do to help protect your crops from pests and diseases. These cultural methods cost nothing and are very effective. Remember, it is better to *prevent* a problem than to solve it:

1 *Keep your crops growing strongly*, by carefully following the methods you have been taught. Try to do everything properly – preparing the ground, manuring and fertilizing, setting out, spacing, weeding, watering etc. Then your plants should be healthy and strong, and have a better chance of resisting pests and diseases.

2 *Use good seed*, 'certified free from disease' by the supplier. Never take your seed from plants that were diseased.

3 *Choose varieties* of crops that grow well in your area.

4 *Keep the garden clean* of diseased plants. This is called **field hygiene**. Remove completely any

plants that should not be there, including weeds, and any that show signs of diseases or pests. This is called **roguing**. Prune dead or diseased leaves or parts of plants. Burn all diseased or infested plant materials, and do not put them on the compost heap. (Why should you not put them on the compost heap?)

5 *Wash your hands* after handling diseased plants. Fungal spores and viruses can be spread from one plant to another on your hands. Pruning tools should be sterilized between pruning one plant and the next. This is important because open wounds are easy entry points for fungal attack.

6 *Practise crop rotation*, to avoid a build up of pest and disease organisms in the soil. Move the crops each year to a new place.

7 *Dig the garden after harvesting* to bury pest organisms, and leave the ground clear of plants. Leave the land to **fallow** or rest for a time. This gives it a chance to recover fertility.

8 *Encourage predators*, such as the ladybird and the praying mantis.

Now a poem – not much to do with the chapter, but just a nice poem!

June Bug

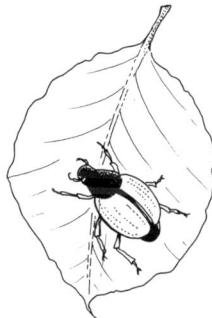

Bug like a coffee bean
Thrown on this tabletop
Beside my paper and pen,
You startle me with your rap.

You, on this hot June night
Which opens window and door,
Come like an intimate
From June of a former year.

Then I, a boy with a book
In a room where a bare bulb glared,
Slept – and struggled awake;
Round me the June bugs whirred.

And, by the inkwell, one
Trundled, a frill of wing
Glinting like cellophane:
On the very lip he clung ...

You're off? No reason to feel
That you, sir, stand on the brink
Of some disastrous fall
Into a pool of ink.

Edward Lucie-Smith.

Questions

1 (a) List four biotic factors.

   (b) Describe any two of these factors that are harmful to plants.

   (c) Describe any two that are helpful to plants.

2 (a) What is a 'weed'?

   (b) Describe three ways in which weeds can harm a growing crop.

   (c) How can weeds be controlled?

3 (a) Explain, with the help of diagrams, how to make a collection of dried, pressed weeds and mount them on paper.

   (b) What information should be written on the paper beside each weed?

4 (a) Give an example of a 'biting and chewing pest'. Describe the damage it does, and how it can be controlled.

   (b) Give an example of a 'piercing and sucking pest'. Describe the damage it does, and how it can be controlled.

5 (a) Explain how micro-organisms in the soil are helpful to plants.

   (b) Give two examples of harmful micro-organisms, and say what damage they cause to plants.

   (c) How can they be controlled?

APPENDIX OF PLANT NAMES

Common name	Scientific name
Aloe	*Aloe barbadensis*
Arrowroot	*Maranta arundinacea*
Bamboo	*Bambusa species*
Banana	*Musa species*
Beans	*Phaseolus vulgaris*
Beet	*Beta vulgaris*
Breadfruit tree	*Artocarpus altilis*
Broom weed	*Sida acuta*
Cabbage	*Brassica oleracea*
Cocoa	*Theobroma cacao*
Coconut	*Cocos nucifera*
Coffee	*Coffea arabica*
Carrot	*Daucus carota*
Cashew	*Anacardium occidentale*
Cassava	*Manihot esculenta*
Cedar	*Cedrus lebani*
Chac chac	*Crotalaria species*
Chilli pepper	*Capsicum frutescens*
Cocoyam	*Colocasia esculenta*
Cotton	*Gossypium species*
Croton	*Codiaeum variegatum*
Dasheen (cocoyam)	*Colocasia esculenta*
Date palm	*Phoenix dactylifera*
Eddo	*Colocasia antiquorum*
Egg plant	*Solanum melongena*
Ginger lily	*Zingiber officinale*
Grapefruit	*Citrus paradisi*
Green pepper	*Capsicum annuum*
Groundnut	*Arachis hypogaea*
Guava	*Psidium guajava*
Hibiscus	*Hibiscus rosa-sinensis*
Hibiscus (variegated)	*Hibiscus acalypta*
Ivy	*Hedera helix*
Lemon	*Citrus limon*
Lime	*Citrus aurantifolia*
Lettuce	*Lactuca sativa*
Maize, corn	*Zea mays*
Mango	*Mangifera indica*
Mangrove (with prop roots)	*Rhizophora mucronata*
Mangrove (with breathing roots)	*Sonneratia alba*
Nutmeg	*Myristica fragrans*
Okra (ochro)	*Hibiscus esculentus*
Onion	*Allium sepa*
Orange	*Citrus sinensis*
Orchids	*Bulbophyllum species*

Common name	Scientific name
Pangola grass	*Digitaria decumbens*
Paspalum	*Paspalum species*
Pawpaw	*Carica papaya*
Pepper (chilli)	*Capsicum frutescens*
Pepper (sweet)	*Capsicum annuum*
Pigeon pea	*Cajanus cajan*
Pineapple	*Ananas comosus*
Plantain	*Musa species*
Pumpkin	*Cucurbita maxima*
Radish	*Raphanus sativus*
Rice	*Oryza sativa*
Rose	*Rosa species*
Runner bean	*Phaseolus multiflorus*
Savannah grass	*Anoxus compressus*
Silk cotton tree	*Ceiba pentandra*
Soya bean	*Glycine max*
Spanish thyme variegatum	*Thymus vulgaris*
Spinach	*Spinacea oleracea*
String bean	*Phaseolus vulgaris*
Sugar cane	*Saccharum officinarum*
Sweet potatoes	*Ipomea batatas*
Teak	*Tectona grandis*
Tobacco	*Nicotina tabacum*
Tomato	*Lycopersicum esculentum*
Wonder of the world	*Bryophyleum*
Yam	*Dioscorea alata*

Some plant families

Solanaceae	Includes tomato, melongene, sweet peppers and hot peppers.
Cruciferae	Includes cabbage, patchoi, mustard, broccoli, kale and cauliflower.
Gramineae	Includes maize (corn), rice, oats, wheat, barley, elephant grass, pangola grass and bamboo. This is the largest plant family.
Leguminosae	Includes peas, beans, lupins, broom and clover.

INDEX